SpaceX's Dragon: America's Next Generation Spacecraft

Erik Seedhouse

SpaceX's Dragon: America's Next Generation Spacecraft

 Springer

Published in association with
Praxis Publishing
Chichester, UK

Erik Seedhouse
Assistant Professor, Commercial Space Operations
Embry-Riddle Aeronautical University
Daytona Beach, Florida

SPRINGER-PRAXIS BOOKS IN SPACE EXPLORATION

Springer Praxis Books
ISBN 978-3-319-21514-3   ISBN 978-3-319-21515-0   (eBook)
DOI 10.1007/978-3-319-21515-0

Library of Congress Control Number: 2015945951

Springer Cham Heidelberg New York Dordrecht London
© Springer International Publishing Switzerland 2016

Cover design: Jim Wilkie
Project copyeditor: Christine Cressy

Printed on acid-free paper

Springer International Publishing AG Switzerland is part of Springer Science+Business Media (www.springer.com)

Contents

Acknowledgments

In writing this book, the author has been fortunate to have had five reviewers who made such positive comments concerning the content of this publication. He is also grateful to Maury Solomon at Springer and to Clive Horwood and his team at Praxis for guiding this book through the publication process. The author also gratefully acknowledges all those who gave permission to use many of the images in this book, especially SpaceX.

The author also expresses his deep appreciation to Christine Cressy, whose attention to detail and patience greatly facilitated the publication of this book, to Jim Wilkie for creating the cover, and to Project Manager, Ms Sasireka, and her team, for guiding this book through the publication process.

About the author

Erik Seedhouse is a Norwegian-Canadian suborbital astronaut whose life-long ambition is to work in space. After completing his first degree in Sports Science, the author joined the legendary 2nd Battalion the Parachute Regiment. During his time in the "Para's", Erik spent six months in Belize, where he was trained in the art of jungle warfare. Later, he spent several months learning the intricacies of desert warfare on the Akamas Range in Cyprus. He made more than 30 jumps from a Hercules C130 aircraft, performed more than 200 abseils from a helicopter, and fired more light anti-tank weapons than he cares to remember!

Upon returning to the comparatively mundane world of academia, the author embarked upon a master's degree in Medical Science. He supported his studies by winning prize money in 100-kilometer running races. After placing third in the World 100km Championships in 1992 and setting the North American 100-kilometer record, Erik turned to ultra-distance triathlon, winning the World Endurance Triathlon Championships in 1995 and 1996. For good measure, he also won the inaugural World Double Ironman Championships in 1995 and the infamous Decatriathlon – an event requiring competitors to swim 38 kilometers, cycle 1,800 kilometers, and run 422 kilometers. Non-stop!

In 1996, Erik pursued his PhD at the German Space Agency's Institute for Space Medicine. While conducting his PhD studies, he found time to win Ultraman Hawai'i and the European Ultraman Championships as well as completing the Race Across America bike race. Due to his success as the world's leading ultra-distance triathlete, Erik was featured in dozens of magazine and television interviews. In 1997, *GQ* magazine nominated him as the "Fittest Man in the World".

In 1999, Erik decided it was time to get a real job. He retired from being a professional triathlete and took a research job at Vancouver's Simon Fraser University. In 2005, the author worked as an astronaut training consultant for Bigelow Aerospace and wrote *Tourists in Space*, a training manual for spaceflight participants. In 2009, he was one of the final 30 candidates in the Canadian Space Agency's Astronaut Recruitment Campaign. Erik works as an astronaut instructor, professional speaker, triathlon coach, and author. Between 2008 and 2013, he served as director of Canada's manned centrifuge and hypobaric operations.

In addition to being a certified suborbital scientist-astronaut, triathlete, centrifuge operator and director, pilot, and author, Erik is an avid mountaineer and is currently pursuing his goal of climbing the Seven Summits. *SpaceX's Dragon: America's Next-Generation Spacecraft* is his nineteenth book. When not writing and training astronauts, he spends as much time as possible in Kona on the Big Island of Hawai'i and in Sandefjord, Norway. Erik is based on Florida's Space Coast and is owned by his rambunctious cat, Lava.

Acronyms

AAU	Animal Access Unit
ACBM	Active Common Berthing Mechanism
AGPS	Assisted Global Positioning System
ALT	Approach and Landing Test
APAS	Androgynous Peripheral Attach System
ARC	Ames Research Center
ASDS	Autonomous Spaceport Drone Ship
ASGSR	American Society for Gravitational and Space Research
ATV	Automated Transfer Vehicle
BEAM	Bigelow Expandable Activity Module
BFR	Big Falcon Rocket
BTNR	Bimodal Thermal Nuclear Rocket
CAL	Cold Atom Lab
CATS	Cloud Aerosol Transport System
CBM	Common Berthing Mechanism
CBR	Certification Baseline Review
CCDev	Commercial Crew Development
CCiCap	Commercial Crew Capability
CCP	Commercial Crew Program
CCtCap	Commercial Crew Transportation Capability
CDR	Critical Design Review
CFD	Computational Fluid Dynamics
COTS	Commercial Orbital Transportation System
CPR	Certification Products Contract
CRS	Commercial Resupply Services
CSA	Canadian Space Agency
CST	Crew Transportation System
CUCU	Commercial UHF Communications Unit
DAV	Descent Assist Vehicle

DDTE	Design, Development, Test, and Evaluation
DRM	Design Reference Mission
EDL	Entry, Descent and Landing
EDS	Emergency Detection System
EOTP	Enhanced ORU Temporary Platform
ERV	Earth Return Vehicle
ETA	Experimental Test Article
FAA	Federal Aviation Administration
FAR	Federal Acquisitions Regulations
FDM	Free Drift Mode
FOR	Flight Operations Review
FRR	Flight Readiness Review
FTA	Flight Test Article
FTS	Flight Termination System
GTO	Geostationary Transfer Orbit
HDEV	High Definition Earth Viewing
HRSI	High-temperature Reusable Surface Insulation
HTV	H-I Transfer Vehicle
ICBM	Intercontinental Ballistic Missile
IRVE	Inflatable Re-entry Vehicle Experiment
IMU	Inertial Measurement Unit
IMMT	ISS Mission Management Team
ISS	International Space Station
JRMS	TEM Remote Manipulating System
KOS	Keep Out Sphere
LAS	Launch Abort System
LCC	Launch Commit Criteria
LEE	Latching End Effector
LEO	Low Earth Orbit
LIDS	Low Impact Docking System
LRT	Launch Readiness Test
MAV	Mars Ascent Vehicle
MCT	Mars Colonial Transporter
MDM	Multiplexer Demultiplexer
MERLIN	Microgravity Experiment Research Locker/INcubator
MLG	Main Landing Gear
MMOD	Micrometeoroid Orbital Debris
MOI	Mars Orbit Insertion
MSL	Mars Science Lander
MPCV	Multi Purpose Crew Vehicle
MTV	Mars Transit Vehicle
NDS	NASA Docking System
NIAC	NASA Institute for Advanced Concepts
NLG	Nose Landing Gear
NSPIRES	NASA Solicitation and Proposal Integrated Review and Evaluation System

NTR	Nuclear Thermal Rocket
OMAS	Orbital Maneuvering and Attitude System
OPALS	Optical Payload for Lasercomm Science
OPF	Orbiter Processing Facility
OV	Orbital Vehicle
PCM	Post Certification Mission
PDR	Preliminary Design Review
PDT	Propulsive Descent Technologies
PICA	Phenolic Impregnated Carbon Ablator
RAD	Radiation Assessment Detector
RCS	Reaction Control System
RSV	Respiratory Syncytial Virus
RTLS	Return To Launch Site
RWS	Robotic Work Station
SAA	Space Act Agreement
SAR	Search and Rescue
SCA	Shuttle Carrier Aircraft
SEI	Space Exploration Initiative
SIGI	Space Integrated GPS/INS
SLC	Space Launch Complex
SLS	Space Launch System
SORR	Stage Operations Readiness Review
SPDM	Special Purpose Dexterous Manipulator
SQU	Space Qualification Unit
SRP	Supersonic Retro-Propulsion
SRR	System Requirements Review
SSIKLOPS	Space Station Integrated Kinetic Launcher for Orbital Payload Systems
SSRMS	Space Station Remote Manipulator System
STP	Supersonic Transition Problem
TDRS	Tracking Data and Relay Satellite
TEI	Trans-Earth Insertion
TIM	Technical Interchange Meeting
TMI	Trans-Mars Injection
TPS	Thermal Protection System
VSE	Vision for Space Exploration
VTVL	Vertical Take-off Vertical Landing
WDR	Wet Dress Rehearsal
ZCG-FU	Zeolite Crystal Growth Furnace Unit

Foreword

In May 2014, SpaceX CEO, Elon Musk, pulled back the curtain on Dragon V2, the spaceship that his commercial spaceflight company hopes will carry NASA astronauts to the International Space Station (ISS) as soon as 2017. The unveiling of the Dragon V2 couldn't have come at a better time. Just a couple of weeks earlier, Russia's deputy prime minister vowed to bar NASA from hitching rides to the ISS aboard Russian Soyuz spacecraft in retaliation for Western sanctions imposed on Russia in response to the Ukraine crisis.

The fortuitous timing – together with Dragon V2's sleek design – will make the futuristic spacecraft a very attractive option for NASA, which is also considering designs by Boeing and Sierra Nevada. But more important to SpaceX is the advance towards the core company objective of reusability. Dragon V2, which was unveiled just weeks after SpaceX demonstrated technologies key to developing a reusable first rocket stage, can be retrieved, refurbished, *and* re-launched. It's a concept with the potential to completely revolutionize the economics of a spaceflight industry where equipment costing hundreds of millions of dollars is often discarded after a single use. With an egg-like shape, soft-white exterior, and SpaceX's name and stylish blue dragon logo emblazoned on its surface, the Dragon V2 spacecraft is much more modern than the Apollo capsules. Inside, the capsule is elegant but sparse, with seven couches facing upward beneath a large, flat-panel display, which serves as the craft's only controls.

SpaceX's Dragon: America's Next-Generation Spacecraft describes the extraordinary feats of engineering and human achievement that have placed this extraordinary spacecraft at the forefront of the launch industry and positioned it as the most likely candidate for transporting humans not only to the ISS, but also to Mars.

1

SpaceX

Elon Musk. Credit: SpaceX

ONE MAN'S MISSION

In 2015, Elon Musk doesn't need to be introduced. Like Wernher von Braun, Musk is a legend in his own lifetime, and deservedly so. A billionaire many times over thanks in part to PayPal and Tesla, Musk is a transformational technologist who has plans to retire on Mars. Why? Not to inspire or to make money. Musk's aim is nothing less than to make us

© Springer International Publishing Switzerland 2016
E. Seedhouse, *SpaceX's Dragon: America's Next Generation Spacecraft*,
Springer Praxis Books, DOI 10.1007/978-3-319-21515-0_1

a multi-planet species. For Musk, colonizing Mars is extinction insurance, and he wants to send a manned mission sooner rather than later. And cheaper. As he points out in interviews, the Curiosity rover that is ambling around on Mars cost more than US$3 billion. Musk reckons, for that money, we should be sending humans and he has a plan to do exactly that. Shortly before SpaceX was founded, Musk remembers a late night spent searching for a NASA website that outlined the plans for a manned mission to Mars. It was 2001 and the Shuttles were still flying at a flight rate that was sufficient to persuade the public that manned spaceflight was still a serious concern. Now, in 2015, with the Shuttle a rapidly receding memory, the notion that manned spaceflight is a vibrant arena is pure delusion. In 2015, American, European, and Canadian astronauts climb on board the Soyuz for their ride to the International Space Station (ISS), and there aren't many people who want to watch spacefarers climb into an aged space vehicle to a destination that is 1,000 times closer to Earth than the Moon. Mars? That isn't even on most people's radar. But many people have accepted this as the norm. A manned mission to Mars? That won't happen for decades, especially if we rely on a government to get us there. A few decades ago, following the Moon landing, the thinking was different. The Apollo program would springboard us to Mars and visits to the Moon and the Red Planet would become routine. Permanent bases would be built on the lunar surface and, after settling on Mars (Figure 1.1), we start shooting for the outer planets. Nothing was impossible. Except that it was.

Back in 2001, when Musk (see sidebar) discovered there were no manned Mars missions on the books, he reckoned the US was no longer interested in manned space exploration. But two years later, the *Columbia* accident proved otherwise. Still, the budget for NASA had been atrophying at an alarming rate since the mid-1960s, when the agency received 4.4% of the federal budget. In the 2000s and 2010s, the NASA budget was just 0.5% of the total US budget, which is part of the reason Musk couldn't find any information about a manned Mars mission. Appalled at the lack of progress, Musk began dreaming up a manned Mars mission of his own, but first he needed to build a rocket company, and so SpaceX was born. The story of this visionary company is told in *SpaceX: Making Commercial Spaceflight a Reality*, which was written by this author and published in 2013, so what follows is a brief overview of how Musk made his company the success that it is today.

1.1 Musk's goal is to colonize Mars. Credit: NASA

Elon Musk

Elon Musk is a genius on a mission. Those who follow the goings on in the world of commercial spaceflight will know all about the Falcon 9 and Dragon, and may assume Musk's primary goal is to ferry astronauts to the ISS, because this is the story the media seem to focus on. But Musk's vision extends far beyond low Earth orbit (LEO). He wants to transport people to Mars. Thousands of them. And a big part of that plan is SpaceX, Musk's rocket company that is based in El Segundo, California. SpaceX's hangar is a hangar like any other, except SpaceX's place of work happens to be the site of a revolution in the way spaceflight is conducted. At the time of writing, SpaceX had been around for just 13 short years but, in that time frame, it has achieved more than most national space programs. It has developed its own rocket engines, launcher, *and* spacecraft, and is now poised to deliver astronauts to the ISS. With an employee roster of little more than 1,000, SpaceX has been created thanks to one man and one man only: Elon Musk.

Having bought his first computer when he was 10, Musk, who lived in South Africa as a youngster, learned how to write commercial software and put that knowledge to good use by writing a space – what else? – game called *Blastar*. He was only 12 years old. At 17, he traveled to Canada before beginning life as a student at the University of Pennsylvania. He gained one degree in physics and one in economics, and then headed to Stanford in 1995, where he spent just 48 hours before deciding it would be more exciting to start an internet company. And so Zip2 was born. Zip2 produced publishing software. Musk sold the company in 1999 for a cool US$300 million and then founded X.com, which eventually became PayPal. Perhaps you've heard of it? Musk sold PayPay to eBay for US$1.5 billion in 2002.

For most people, this would be enough. Sit back, head for Tahiti, and soak up your success in the sunshine. But not Musk. Becoming a billionaire was just the beginning because, besides being extremely rich, Musk is also very, *very* single-minded. And one of his goals was to make space affordable and along the way send people to Mars. Never mind that these goals have eluded governments for decades. Musk reckoned the government's way of doing business in space was inefficient in the extreme and he decided he would do it more affordably. And that's what he did. His rockets place cargo into space at a fraction of the cost of his rivals and his company's development of Dragon has been lightning fast.

Inevitably, all this success has cast Musk in the media spotlight. Not surprising really, given that Musk was the inspiration for *Iron Man*'s Tony Stark (Musk made a cameo appearance in *Iron Man 2*), but Musk would rather spend his time dreaming up how he can get us to Mars than answering questions for a magazine. After all, this is a man who spends 100 hours a week at work and logging so many hours means there just isn't time to indulge the media or spend time telling people about his private life (Musk has neither a Facebook nor a Twitter account). While he is a businessman and a celebrity, the word that most accurately defines Musk is "dreamer." He didn't create SpaceX to make another tonne of money. He created it as a stepping stone to help humanity on their way to Mars. He reckons we can get there in 20 years.

When SpaceX launched in 2002, Musk was trying to buy repurposed intercontinental ballistic missiles (ICBMs) for just US$7 million each. He had a number of meetings with the Russians and was close to purchasing three ICBMs for US$21 million, only to be told that the Russians meant US$21 million for *one*. Fed up of dealing with the Russians, Musk decided to build his own rocket. The challenges facing Musk were intimidating. Between 1957 and 1966, the US had sent 429 rockets into orbit, a quarter of which failed. And they had done this with nearly unlimited budgets. Undeterred, Musk started recruiting veterans from the aerospace world in El Segundo, California. One of his first employees was Tom Mueller, a leading propulsion expert. Mueller was followed by more rocket engineers joining the SpaceX cause. To run the day-to-day operations of his company, Musk hired Gwynne Shotwell (Figure 1.2), who became SpaceX's seventh employee as Vice President of Business Development (today Shotwell is President).

1.2 Gwynne Shotwell, SpaceX's President and Chief Operating Officer. Before being recruited by Elon Musk, Shotwell worked for the Aerospace Corporation, where she managed a study for the Federal Aviation Administration on the subject of commercial space transportation. Credit: SpaceX

As he was building his team, Musk got down to the business of building rockets and rocket engines. First was the Falcon 1, a two-stage rocket designed to lift 570 kilograms into LEO. But getting that payload into orbit proved a headache. After being kicked out of the Complex 3 West launch site at Vandenberg, Musk moved the launch to an alternate launch site on Omelek Island in the Kwajalein Atoll (part of the Republic of the Marshall Islands). Even though Falcon 1 had not flown a single mission, Musk had already signed three launch contracts and had sunk US$100 million into SpaceX. There was a lot riding on the first launch. The US$6.7 million, 21-meter-tall Falcon 1 was the first in a family of boosters planned by SpaceX to offer a more affordable way to launch satellites. The vehicle's first launch attempt came on 26 November 2005, but the launch was scrubbed. The second launch attempt took place on 19 December 2005, but a faulty valve resulted in another scrub. The third launch attempt on 10 February 2006 also resulted in a scrub after problems occurred during a planned engine test.

The fourth launch attempt was set for 25 March 2006. This time, everything went by the book – until T+26 seconds, when the Falcon 1 rapidly pitched over and crashed 15 seconds later. Four years of work; millions *and* millions of dollars; endless 80-hour-plus work weeks – it was a major setback. But SpaceX went back to work, eventually making 112 changes to the rocket and the launch sequence. With the changes made, a 20 March 2007 launch date was set for attempt number five. After a minor glitch following an automatic abort, Falcon 1 rose from the launch pad and accelerated out of sight. In El Segundo, the champagne flowed. Five minutes into flight, the rocket's second stage started to spin and after flame-out it continued to roll. At T+11 minutes and 11 seconds, the video feed went blank. Despite the anomaly, Musk claimed success, since all major milestones had been met. With the successful launch of Falcon 1, Musk could get back to the business of further reducing launch costs and making plans for that ultimate goal: Mars.

Musk's goal was to increase the reliability of space access by a factor of 10. It was a bold aim, but Musk reckoned it could be achieved by eliminating the traditional layers of management internally and sub-contractors externally, which is what Musk did. By keeping most of the manufacturing in-house, SpaceX reduced its costs, kept tighter control of quality, and ensured a tight feedback loop between the design and manufacturing teams. Thanks to this scaling-down approach, SpaceX won a 2008 US$1.6 billion contract to ferry cargo to the ISS. Another US$118 million followed in 2010 to help SpaceX complete its original demonstration agreement. Then, in 2011, with the retirement of the Shuttle, NASA turned to the commercial sector to transport its astronauts to the orbiting outpost. All of a sudden, Musk found himself in a position of partnership and power. For a modest investment, NASA kick-started a new industry, giving companies like SpaceX a chance to demonstrate that it could go to space more efficiently and more affordably than old-line government contractors. And SpaceX made good on that investment by delivering its first Dragon spacecraft into LEO in 2010. Then, in May 2012, Dragon docked with the ISS for the first time and, in October of that year, it completed the first resupply run on SpaceX's US$1.6 billion NASA contract.

In September 2014, NASA announced SpaceX had been awarded a US$2.6 billion Commercial Crew Transportation Capability (CCtCap) contract for the development of spacecraft to transport astronauts to the ISS, with a start date of 2017. For SpaceX, that spacecraft is Dragon (Figure 1.3). In 2015, Musk is gearing up to put humans in space,

1.3 A very early Dragon photographed in 2008. This pressure vessel was used in factory tests. Credit: SpaceX

which brings him one step closer to his ultimate goal: Mars. It's an amazing trajectory considering that, less than 10 years previously, Musk had been on the verge of bankruptcy and the first three Falcon 1 launches had failed. Aerospace hacks had sharpened their pens and written him off, yet here he is preparing for humanity's best shot of getting us to Mars. Sure, SpaceX has had its share of delayed launches and the occasional poorly timed mishap, but Musk is getting a reputation for doing commercial spaceflight better than anyone else. And once his company starts ferrying astronauts to the ISS, he will be on the cusp of preparing to do what no one has done for nearly 50 years: transporting humans to a new world.

> "There needs to be an intersection of the set of people who wish to go, and the set of people who can afford to go. And that intersection of sets has to be enough to establish a self-sustaining civilisation. My rough guess is that for a half-million dollars, there are enough people that could afford to go and would want to go. But it's not going to be a vacation jaunt. It's going to be saving up all your money and selling all your stuff, like when people moved to the early American colonies."
>
> *Elon Musk, interview with* Aon Magazine, *30 September 2014*

Colonizing Mars

Even when Mars is at closest approach, the Red Planet is 150 times farther than the Moon. Over the decades, dozens of manned Mars mission architectures have been suggested, all of which have been obscenely expensive. But Musk reckons he can charge his passengers (colonists) just US$500,000. Let's put that number into perspective. A space-tourism company by the name of Space Adventures is charging US$150 million a seat for a round-trip ticket to the Moon sometime in 2017. Assuming all goes to plan, the mission will launch from Russia and the tourist and pilot will stop off at the ISS before continuing on to the Moon. The length of the trip is expected to be 17 days. That's a long time to be stuck in a Soyuz! And that US$150 million doesn't include landing. Yet Musk reckons he can budget for US$500,000 a ticket to Mars, landing included. What does he have in mind? Well, we'll discuss this later in the book. For now, let's just say I'm not betting against him. While Musk has a vehicle in mind (reusable of course), the real challenge will be making the numbers work. Of course, one way to do this is reusability. Traditionally, rockets are thrown away after every flight, which is a terrible way to do business, especially if you plan to send 80,000 people to a distant planet. But Musk reckons he can fix the problem of reusability and, by doing so, drive down mission costs to just tens of dollars per kilogram to LEO. If he does that, then he'll be in a good position to charge budget prices for seats on board his colony spaceship.

Eighty thousand people on Mars sounds like science fiction to most people, especially when you consider no government agency has even come close to sending just one astronaut there. But, until a few years ago, a commercial assembly line for spacecraft didn't exist. Don't forget, we're talking about a visionary – someone who pushed way beyond the envelope to achieve what many thought was downright impossible. And let's not forget, Musk wasn't the first to achieve the impossible. Remember Wernher von Braun? In World War II, the great German was developing V-2 rockets for the Fuhrer, but von Braun's dream was to use rockets to carry artificial satellites into orbit. Since his superiors weren't too enthused with the idea, von Braun siphoned resources from the weapons program to pursue spaceflight research while making just enough headway on the V-2 to remain credible with his bosses. When it became clear his commanders planned to kill the V-2 scientists to prevent the Allies from acquiring the expertise, von Braun fled to the US.

> "The thing that got me to start Space X was being disappointed that we'd not made progress beyond Apollo. There was this incredible dream of exploration that was ignited with Apollo and it had felt as though the dream had died. And year after year, we did not see improvements in rocket technology. Even before I started SpaceX, my goal was to increase the NASA budget to make that happen. But, as I learned more, I discovered that, unless we improve our rocket technology, it's just not going to matter. Eventually there will be something that happens on Earth, either man-made or natural calamity … to cause the end of civilization."
>
> *Elon Musk being interviewed (FastcoCreate.com), 14 June 2014*

Wernher von Braun

As early as 1948, von Braun was calculating how to get to Mars and, when he wasn't busy working out how to get humans to the Red Planet, he championed the idea of spaceflight, but the Americans weren't interested. Undeterred, von Braun continued his work on a mission to Mars, the outcome of which was presented on a Walt Disney show to an audience of 42 million viewers. Like Musk, von Braun dreamed big. His Mars mission would send 70 astronauts on 10 spacecraft, with each vehicle weighing 3,720 tonnes! Thinking way ahead of his time (this was the 1950s), von Braun also envisaged a fully reusable launch vehicle that would deliver 25 tonnes of cargo and 14.5 tonnes of fuel to LEO. Von Braun's Projekt Mars never got off the ground. In 1957, Sputnik caught the attention of the US and von Braun was tasked with developing a plan to land Americans on the Moon. After outlining his proposal for landing men on the Moon, the Americans were all ears, and so the Space Race was born. As the creator of the Saturn V, von Braun was the first true visionary of the American space program. It was von Braun's creative genius that propelled the US to the Moon: without him, there is a good chance Kennedy's dream of reaching the Moon before the end of 1969 would have withered and died.

In the short history of manned spaceflight, many of most revolutionary advances have come not from the dictates of a politician, or the incrementalism of a bureaucrat, but in the compelling power of a visionary in the von Braun mould: to redirect missions to their cause, and convert those sitting on the fence to their way of thinking. Fast forward 50 years and we have another visionary in the shape of Elon Musk. Musk has succeeded because he didn't wait for the government to deliver the perfect program – he started with the tools at hand and moved forward. Aggressively. One of Musk's greatest qualities is his extraordinary ability to believe in his own vision to such an extent that he doesn't really think that what's he's trying to do is particularly risky. What he's trying to do is nothing less than open up an interplanetary era. It's real Tony Stark stuff. And part of the reason he will succeed is because he is willing to take risks. In the juggernaut industry that is the aerospace world, there is an extraordinary aversion to risk. Even when better technology is available, aerospace companies still prefer to use legacy components, some of which were designed way back in the 1960s. Take Orbital Sciences as an example. This company, like SpaceX, has a contract to ferry cargo to the ISS, but their Antares launcher uses Russian rocket engines (the AJ-26) from the 1960s. On 28 October 2014, Orbital Sciences suffered a launch failure, and the finger of suspicion was pointed at that Russian rocket engine. So what did Orbital Sciences do? They contracted with the Russians to buy another rocket engine (the RD-181) – 60 of them, at a cost of US$1 billion! SpaceX doesn't go down this road of outsourcing: no subcontracting or sub-subcontracting for this company, which means their costs are streamlined and rocket launches have been brought down by a factor of 10.

2

Commercial Crew Programs

NASA's Commercial Crew Program. Credit: NASA

© Springer International Publishing Switzerland 2016
E. Seedhouse, *SpaceX's Dragon: America's Next Generation Spacecraft*,
Springer Praxis Books, DOI 10.1007/978-3-319-21515-0_2

"SpaceX is deeply honoured by the trust NASA has placed in us. We welcome today's decision and the mission it advances with gratitude and seriousness of purpose."

SpaceX Chief Executive and chief designer Elon Musk, 16 September 2014

In January 2015, SpaceX's Dragon moved one step closer to flying astronauts to the International Space Station (ISS) with the announcement that the vehicle had passed NASA's Certification Baseline Review (CBR) which was an assessment of how SpaceX planned to transport crews to and from the orbiting outpost using Dragon V2. The CBR was part of SpaceX's Commercial Crew Transportation Capability (CCtCap) contract with NASA and CCtCap was one element of the agency's Commercial Crew Program (CCP) (Figure 2.1). But what is the CCtCap, how do all the other CCP-related acronyms fit into the commercial funding puzzle, and why is it important to know this anyway? Well, it's important because, although Dragon, Red Dragon, and Dragon V2 are commercial vehicles, much of the design and development of these spacecraft, which we'll get to in the next chapter, is funded by NASA dollars. So what follows is a sort of CCP 101.

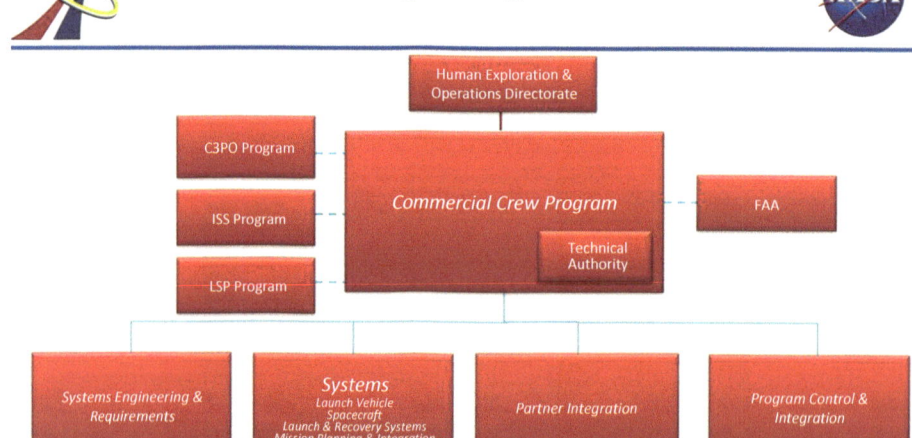

2.1 NASA's Commercial Crew Program (CCP) structure. Credit: NASA

COMMERCIAL CREW PROGRAM (CCP): A PRIMER

The catalyst for the CCP can be traced back to the *Columbia* accident in 2003. The second Shuttle catastrophe was the trigger for President Bush to order NASA to complete the ISS and do away with the Orbiter by the end of 2010. Instead of flying Space Shuttles, NASA was told to develop new vehicles for missions to the Moon and to set up research bases on the lunar surface. NASA knuckled down and came up with the Constellation Program and began to develop a monster super rocket to launch lunar landers and habitats to the Moon. The Constellation Program, like so many space programs, was short-lived, because, when President Obama was voted in, one of his first acts was to kill Constellation. The one element that survived the cull was the Orion capsule, and it became the centerpiece for what has since become known as the flexible path architecture. Part of this recalibration includes the development of a monstrous Space Launch System (SLS) to launch the Orion on deep-space missions: asteroids have been suggested, as have missions to Mars, but the truth is nobody really knows for sure. The first unmanned test flight of Orion took place on 5 December 2014 and the first crewed mission is expected sometime in 2021. Of course, that date doesn't do American astronauts much good because they are still reliant on the aging

2.2   The most frequently used and most reliable spacecraft ever: the Soyuz. It also happens to be a little pricey, with seats costing upwards of US$72 million. It is also the only mode of transport to the International Space Station (ISS) if you happen to be a NASA, Canadian, Japanese, or European astronaut. But Dragon may be taking over transportation duties as soon as 2017. Credit: NASA

and progressively more expensive Soyuz (Figure 2.2) for their flights to the ISS. Average seat price: US$72 million! So Obama decided to go the commercial route and encouraged NASA to outsource contracts to the private sector. This was a seismic change for NASA which was used to doing business the NASA way. The NASA way meant that the agency made all the design and development decisions and, at the end of the day, they owned and operated the hardware. But going the commercial route would mean the agency would provide funding and their know-how to the companies who would be responsible for designing and developing new launch vehicles and spacecraft.

The goal of the CCP was very simple: to develop an American commercial crew space transportation system to ensure safe, reliable, and affordable access to and from the ISS and low Earth orbit (LEO). From its inception, the plan was for the CCP to keep on providing funding to commercial companies until one or more had developed a transportation system that met NASA requirements. Once that objective was achieved, one or more companies would fly astronauts to the ISS. The CCP also had a dual benefit because all the money pouring into the commercial space industry couldn't help but spur economic growth and also create new markets. Since the CCP program began, it has awarded more than US$8.2 billion in various contracts up to 2015, and the results have been promising.

Before delving into the various types of contract, it is instructive to review the original approach of developing a manned transportation system and compare this with the commercial way of doing things. At NASA, the agency would define requirements for the vehicle that was needed and then hand that information over to the engineers, who would then develop all the myriad elements of the spacecraft together with all the support systems and operations. Once the design was rubber-stamped, the design was handed over to aerospace contractors, which were usually strategically located around the country. These aerospace contractors then went about the business of actually building the vehicle to NASA's exact specifications. In parallel with this activity, NASA personnel were involved in the testing and processing of the vehicle as it took shape. Then, once the vehicle was ready, NASA personnel started the business of launching and operation of the system until the agency was satisfied the system was safe and reliable. That, in a nutshell, is how every American manned spacecraft (Figure 2.3) has been built. Until now.

The commercial route to building a manned spacecraft is one defined by a process of collaboration. The commercial companies still go through the same steps as NASA did, but in this case it is the commercial company doing the designing and testing, although they do this in close partnership with NASA. It's a little like a group effort. Although the commercial company is free to design the transportation system however they see fit, at the end of the day, the system must meet the agency's rigid requirements. If the requirements are not met, then it's back to the drawing board because there is no money paid out until everything is checked off against NASA's requirements list. Thanks to the collaborative nature of the process, this doesn't happen often.

2.3 The Shuttle Atlantis. The Shuttle was born in 1968 to provide the US with a reusable means of transporting crews to and from a space station. "Cheap" and "routine" were words often bandied about when discussing this revolutionary spacecraft. After North American Rockwell and McDonnell Douglas were granted contracts, NASA set a date of 1977 for operational service. Then things got complicated. Major design modifications resulted in a huge price increase and then development difficulties with the Space Shuttle Main Engines caused more delays. Just five years after its maiden flight in 1981, the program ground to a halt following the Challenger disaster. More modifications and upgrades were called for. The beginning of the end was the Columbia accident in 2002. The cost? For the 30-year service life adjusted for inflation: US$196 billion, which equates to a cost of about US$450 million per flight. Impressive space vehicle? Absolutely. Cheap and routine. Anything but. Credit: NASA

CONTRACTS

NASA wouldn't be NASA without acronyms, and the CCP generated a few more to the long list of abbreviations. One of the new acronyms was SAA, or Space Act Agreement. This was designed with the purpose of defining when an agreement had been made between NASA and the commercial company in question. Contained within a typical SAA were statements relating to milestones, investment, design, and capabilities of the vehicle, and an overview of the technical support required. The main reason for a SAA was to use it as a means to partner with commercial companies, which leads us to the Commercial Crew Development (CCDev) program.

CCDev was a space technology development program, the intent of which was to stimulate development of commercially operated crew vehicles. The program's goal was to select at least two providers to ferry astronauts to the ISS by 2017. Unlike traditional space industry contractor funding, CCDev funding could only be used to fund specific subsystem technology development objectives that NASA wanted for NASA purposes. For example, in 2011, as part of CCDev2, NASA awarded SpaceX US$75 million to develop its launch escape system. In CCDev's first phase (CCDev1), NASA handed out US$50 million to five companies to generate research and development into commercial manned spaceflight concepts and technologies. Later in the program, a second set of CCDev proposals was solicited by NASA for technology development project durations. For example, one of the proposals selected included Blue Origin, which was awarded funding to develop a "pusher" Launch Abort System (LAS), while Boeing received funding to develop its Crew Space Transportation (CST)-100 vehicle.

Following CCDev1, NASA announced it would provide almost US$270 million of funding to four companies contingent on meeting CCDev2 objectives. One of these objectives was the capability of a spacecraft to deliver and return four astronauts and their cargo to and from the ISS while another objective was to prove the capability of providing crew return in the event of an emergency. In this round of funding, SpaceX was awarded that aforementioned US$75 million to develop its LAS for Dragon. The following list gives you an example of the work completed by SpaceX under CCDev2 and the chart (Figure 2.4) provides an outline of the milestones set by NASA.

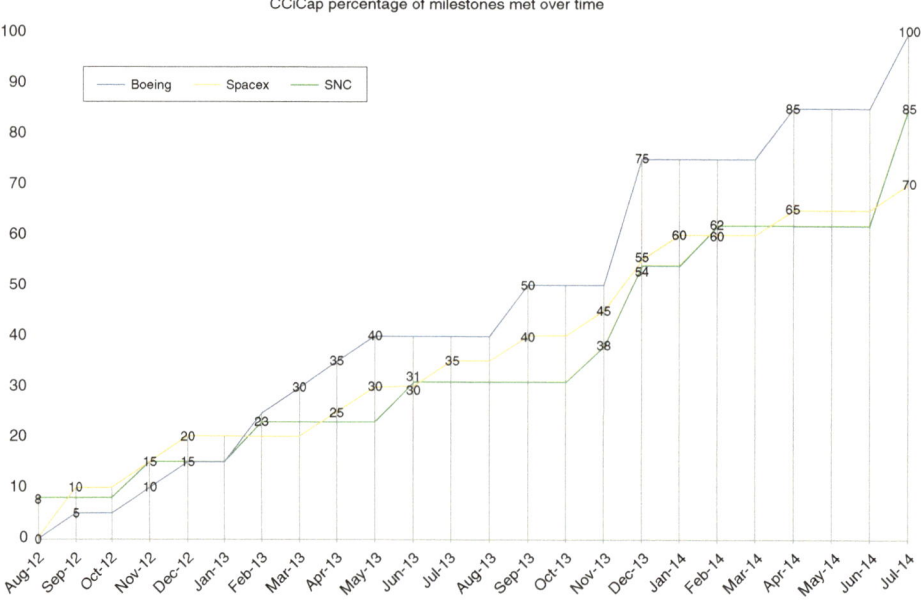

2.4 Commercial Crew Development (CCDev) chart. Credit: NASA

SpaceX summary of work completed during CCDev2

- Launch abort system design, development, and test
- SuperDraco engine: demonstration engine and development engine
- SuperDraco test stand
- Propellant tank
- System components
- Crew systems initial design and development
- Structures including seats and cabin layout
- Environmental control and life support
- Spacesuits
- Launch pad modifications
- Mission operations and recovery
- Crew displays and controls
- Concept of operations
- Abort and other analyses
- Guidance, navigation, and control
- Aerodynamics
- Environments
- Safety and mission assurance
- Human certification plan
- Technical budgets and key performance parameters
- Requirements compliance
- Crew trials.

Following on from CCDev2 was the Commercial Crew Integrated Capability (CCiCap), which was announced on 3 August 2012. Under CCiCap, NASA made funding awards to three commercial companies under signed SAAs. The CCiCap partners were Sierra Nevada Corporation, which received US$212.5 million; SpaceX, which received US$440 million; and the Boeing Company, which received US$460 million. For the CCiCap base period, Space X was tasked with focusing on the following:

1. Completing the integrated design
2. Hardware testing to reduce risk
3. Ensuring crew safety
4. Preparing for NASA certification.

> "This is a decisive milestone in human spaceflight and sets an exciting course for the next phase of American space exploration. SpaceX, along with our partners at NASA, will continue to push the boundaries of space technology to develop the safest, most advanced crew vehicle ever flown."
> *Elon Musk, following the CCiCap announcement, August 2012*

As with all funding programs, SpaceX was required to meet certain milestones, which are listed below. Only if SpaceX met the milestones would it receive the funding.

1. 29 August 2012: CCiCap technical baseline review
2. 30 August 2012: Financial and business review

3. 29 October 2012: Integrated system requirements review
4. 13 December 2012: Ground systems and ascent preliminary design review
5. 28 March 2013: Pad abort test review
6. 7 May 2013: Human certification plan review
7. 18 July 2013: On-orbit and entry preliminary design review
8. 17 September 2013: In-flight abort test review
9. 29–30 October 2013: Safety review
10. 18 November 2013: Flight review of upgraded Falcon 9
11. 20 December 2013: Parachute drop test
12. Integrated critical design review
13. Pad abort test
14. Dragon primary structure qualification
15. In-flight abort test.

For those interested in the finer details of the CCiCap, Table 2.1 outlines the funding as it is tied to the milestones.

The next step in the CCP was CCtCap. Under this funding program, SpaceX won US$2.6 billion to press ahead to develop its Dragon crew vehicle. CCtCap was designed as a two-phase certification program to build and operate a manned vehicle. Two contracts, one to Boeing and one to SpaceX, were awarded in September 2014 following an open competition (Table 2.2).

Who will be first to provide the orbital taxi service? It's difficult to say because Boeing is rather media-shy when it comes to announcing latest developments of their CST-100 vehicle. What we do know is that the CCtCap funding will culminate with a minimum of one test flight carrying an astronaut and both companies will be required to demonstrate that their operations and systems work in space. For NASA, a commercial manned capability can't come soon enough. Assuming SpaceX or Boeing remains on

Table 2.1 SpaceX CCiCap milestone progress.[1]

	% of total award (US$ million)	Cum. % of total award	Milestone	Completed
M1: CCiCap kickoff meeting	40.0	9%	9%	Sep. 2012
M2: Financial and business review	20.0	4%	13%	Sep. 2012
M3: Integrated system requirements review	50.0	11%	24%	Nov. 2012
M4: Ground systems and ascent preliminary design review	35.0	8%	32%	Dec. 2012
M5: Pad abort test review	20.0	4%	36%	Apr. 2013
M6: Human certification plan review	50.0	11%	47%	May 2013
M7: On-orbit and entry preliminary design review	34.0	7%	54%	Jul. 2013
M8: In-flight abort test review	10.0	2%	56%	Sep. 2013
M9: Safety review	50.0	11%	67%	Nov. 2013
M10: Flight review of upgraded Falcon 9	0.0	0%	67%	Dec. 2013
M15A: Dragon parachute tests phase I	15.0	3%	70%	Dec. 2013
M15A: Dragon parachute tests phase II	5.0	1%	72%	Jan. 2014

[1]Source: February 2014 ROI report.

Table 2.2 NASA's source selection explained.

	Boeing	SpaceX	Sierra Nevada
Mission suitability	913 points	849 points	829 points
Technical (525 points)	Excellent	Very Good	Very Good
	488 points	457 points	420 points
Management (400 points)	Excellent	Very Good	Very Good
	372 points	344 points	356 points
Small business (75 points)	Good	Good	Good
	53 points	48 points	53 points
Past performance	V High Confidence	High Confidence	High Confidence
Price	US$3.01 billion	US$1.75 billion	US$2.55 billion

track to fly astronauts to the ISS by 2017, the US will have waited six long years to see an end to the sole source reliance on the Russians.

"It's an incredible testament to American ingenuity and know-how, and an extraordinary validation of the vision we laid out just a few years ago as we prepared for the long-planned retirement of the Space Shuttle. This work is part of a vital strategy to equip our nation with the technologies for the future and inspire a new generation of explorers to take the next giant leap for America. We have been working overtime to get Americans back to space from US soil and end US reliance on Russia. My job is to ensure we get Americans back to space as soon as possible and safely."

NASA Administrator and four-time Shuttle astronaut, Charlie Bolden, during the briefing at the agency's Johnson Space Center in Houston, 26 January 2015

Assuming a Dragon begins ferrying astronauts to the ISS in 2017, will the CCP have been worth it? It depends on how you view success I guess. One of the goals of the CCP was to develop a competitive commercial spaceflight industry and one of the ways to achieve that was to move away from a sole provider because by doing this you get a healthier and more affordable transportation system. With Soyuz seat prices going through the roof, NASA wanted to limit the time it had to keep on writing checks to the Russians and it wanted to stop its reliance on just one system. With Dragon and the CST-100 on the way to being man-rated, you have to say that CCP has been successful.

Of course, the goal of the CCP wasn't just to ignite competition, but to maintain it. If competitiveness can't be maintained, then CCP may be viewed as having failed. One way of assessing the value of CCP is using the lifetime of ISS as a baseline. We know ISS operations will continue to at least 2024. Between the potential start of commercial manned transportation to the ISS and 2024, it is possible that the orbiting outpost's crew could be increased to 14. That is because, in 2015, there are two Soyuz lifeboats and each Soyuz can only carry three astronauts. But, with the seven-crewmember capability of Dragon (and CST-100), it will be possible to evacuate 14. Obviously, an increase from six crewmembers to 14 would dramatically increase the scientific return from the ISS, so in that regard the CCP would be considered a success. Also, since the number of astronauts per Dragon will affect cost, there is a good chance that seat prices will go down over time.

2.5 Bigelow Expandable Activity Module, or BEAM. One day, Dragons could be flying Mr. Bigelow's sovereign customers to these modules. Credit: NASA

But of course the ISS won't be the only destination in LEO. Bigelow has plans to orbit his Bigelow Expandable Activity Modules (BEAMs) (Figure 2.5), which opens up another market for Dragon. In fact, Bigelow has had his inflatable stations ready for some time now but has had to play the waiting game for a commercial crew vehicle to become operational. There are also plans in the works for orbital refueling, satellite servicing, asteroid deflection capabilities, and space tourism, and having manned orbital access is an enabler for these services. As of mid-2015, mission planners – the Flight Planning Integration Panel (FTIP) – have made plans for four commercial crew demonstration flights to the ISS beginning as early as December 2016. If this flight happens to be a flight of Dragon 2, then a second commercial manned flight is set to fly in April 2017. In fact, there is a possibility that there could be two Dragons berthed at the ISS at that time because the April 2017 date overlaps a planned cargo mission: SpX-14. Assuming these dates are met, the collaboration fostered by CCP will have set the US on the path towards regaining independence for its manned spaceflight needs. By that mark, CCP will be seen as a major evolution in the way astronauts are ferried to and from orbit – as long as the funding is approved.

FUNDING

The problem with funding is that Congress isn't approving all the funding that NASA has been requesting to keep the space taxi venture on schedule. In 2014, the Obama administration requested US$1.244 billion to be set aside for the commercial spaceflight initiative, but this was US$439 million more than Congress approved for the budget. But NASA requires the full US$1.244 billion if SpaceX and/or Boeing are to meet their milestones by the dates set in the CCtCap. If the funding isn't forthcoming, then schedules slip and the CCP is short-changed. Not only that, but any delay would result in more money being siphoned off to the Russians to pay for more expensive Soyuz seats. And, if that happens, NASA would have to go back to SpaceX and negotiate and probably recalibrate the milestones. To give you an idea of the disparities of funding requests against funding granted, years 2011 through 2015 are listed below:

- 2011: US$321 million appropriated vs. US$500 million requested;
- 2012: US$397 million appropriated vs. US$850 million requested;
- 2013: US$525 million appropriated vs. US$830 million requested;
- 2014: US$696 million appropriated vs. US$821 million requested;
- 2015: US$805 million appropriated vs. US$848 million requested.

Another impact of funding not being available is the snowball effect of the delay to operational commercial crew services because NASA would have to divert funds for Soyuz seats from their exploration budget. This would mean that plans to send the Orion to an asteroid or some other deep-space destination would be put on hold. It would also probably restrict funding to NASA's other flagship projects such as the SLS and the development of the Orion. But, as always, the agency's program goals are heavily influenced by what goes on in Congress. While buying more Soyuz seats represents a contingency, it is a backup scenario that should be avoided if at all possible, so the hope is that, by 2017, we will see Dragon ferrying its first human cargo to the ISS. The following chapters explain how.

3

Dragon Design, Development, and Test

The first Dragon. Credit: NASA

DESIGNING DRAGON

The business of designing Dragon started at the tail end of 2004. At the time, SpaceX had decided it would pursue development of the capsule with its own funds, but that changed in 2005 when NASA announced its Commercial Orbital Transportation Services (COTS) development program and advertised that it was soliciting proposals for a commercial cargo vehicle which would replace the Shuttle. SpaceX submitted Dragon as its proposal in March 2006. Six months later, NASA awarded SpaceX funding to develop a cargo service to the International Space Station (ISS), the plan being to fund three demonstration flights that

© Springer International Publishing Switzerland 2016
E. Seedhouse, *SpaceX's Dragon: America's Next Generation Spacecraft*,
Springer Praxis Books, DOI 10.1007/978-3-319-21515-0_3

3.1   Dragon in orbit. Credit: SpaceX

would be flown between 2008 and 2010. For this task, SpaceX was awarded US$278 million, provided they met all of NASA's milestones.[1] Because Dragon's (Figure 3.1) primary role was to ferry crews to and from ISS, the capsule didn't need to be as roomy as NASA's five-meter-diameter Orion (Figure 3.2): in contrast, the gum-drop-shaped Dragon measured just 3.6 meters across. Visually, the vehicle bears more than a striking resemblance to the Apollo (Figure 3.3) and Orion capsules, and it sports many of the same features [1].

Heat shield

As you can see in the images of the Orion, Dragon, and Apollo spacecraft, each has almost an identical blunt-cone capsule shape. In the case of Dragon, a nose-cone cap jettisons after launch and the trunk is equipped with two solar arrays. Its heat shield is a proprietary variant of NASA's phenolic impregnated carbon ablator (PICA) material, which is worth further discussion. The main goal of Dragon is to ferry crew and cargo to and from the ISS and to accomplish this safely means a really tough and really reliable heat shield is needed. It also helps the business case if the heat shield can be reused after each flight – remember the palaver following each Shuttle flight when technicians had to check all those tiles? This is part of the reason that SpaceX chose an ablator for its Dragon (another reason is there aren't many thermal protection systems (TPS) to choose from: if you're in the business of designing and building spacecraft, you can either choose the Shuttle's TPS or an

---

[1]   Two years later, on 23 December 2008, NASA awarded a US$1.6 billion CRS contract to SpaceX, with options to increase the contract value to US$3.1 billion. The contract called for 12 cargo flights to the ISS.

3.2   NASA's Orion capsule. Credit: NASA

3.3   Apollo 17. Credit: NASA

ablation-based TPS).[2] Thanks to successes of such high-profile missions as NASA's Mars Science Laboratory (MSL), ablative [2] has been the way to go in recent years, but what is it about ablation that makes it so attractive? We'll answer that, but first let's begin with the definition of ablation: this is simply the decay of the spacecraft's outer layer of skin caused by the tremendous heating due to re-entry speeds. This decay occurs due to convection, which in turn is caused by a pyrolyzing layer that diffuses in the direction of the heated area of the shield. Here, a boundary layer is created by the pressure of re-entry and it is this layer that ensures the heat is redirected away from the shield. What is left is char – basic but effective, and safer than the Shuttle's TPS.

While the Shuttle's TPS was very effective, the one time it failed caused the destruction of the vehicle and the death of seven crew. And 1 out of 134 is just not good enough for commercial spaceflight, or any manned spaceflight endeavor for that matter. Not only that, but maintaining the Shuttle's TPS was time-consuming in the extreme, since each – there were 20,548 high-temperature reusable surface insulation (HRSI) tiles alone – had to be checked prior to each flight. So ablative was the default option, but why PICA? There are number of reasons. First, using an ablative system is efficient, because having engineers spend weeks and weeks checking thousands *and* thousands of tiles is extremely time-consuming. Secondly, ablatives ensure faster re-entries, because PICA (Figure 3.4) can withstand very high temperatures – five times higher temperatures (2,760°C) than the

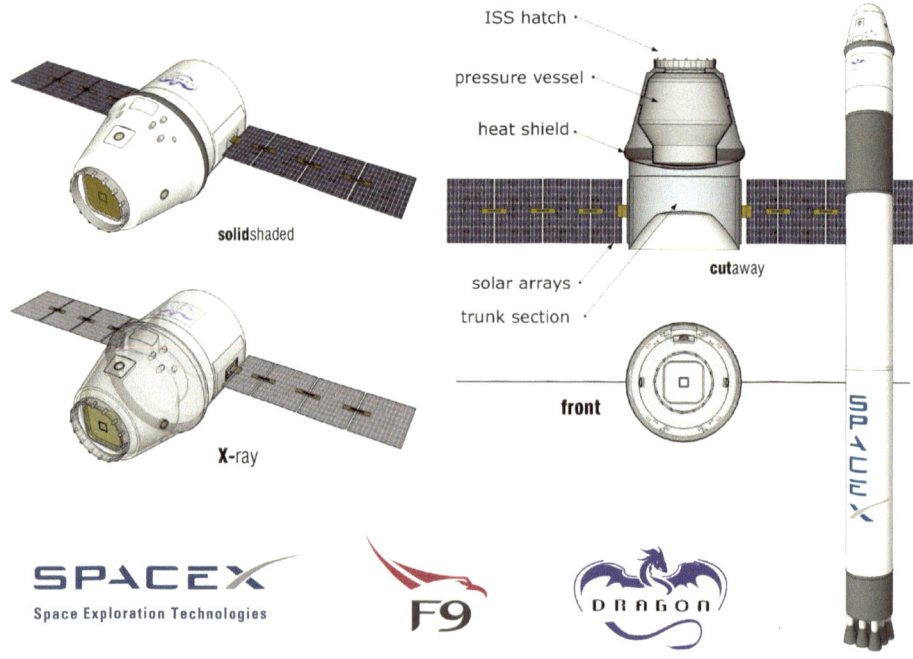

3.4 Dragon's heat shield. Credit: NASA

---

[2]   There is a TPS known as the Inflatable Re-entry Vehicle Experiment (IRVE), but this is in the experimental stage.

3.5 NASA's Stardust return capsule. Stardust was launched in 1999 on a mission to collect dust samples from a comet coma (Wild 2). In 2006, the sample return capsule disengaged from Stardust and re-entered Earth's atmosphere at 12.9 kilometers per second (about Mach 36). Thanks to its PICA heat shield, which reached 2,900°C, the capsule was none the worse for wear. Credit: NASA

Shuttle's tiles [3]. And thirdly, ablative systems ensure ease of production. Yet another attractive feature of PICA is that the material is very, *very* light – only slightly heavier than balsa wood! And for those of you who keep an eye on space records, you will know that PICA has enabled the fastest re-entry – 12.9 kilometers per second – ever, when NASA's Stardust (Figure 3.5) capsule re-entered Earth's atmosphere in January 2006. Despite all these advantages, there was one problem that SpaceX faced when opting to use PICA as Dragon's heat shield: cost. SpaceX wasn't prepared to pay the price that ARA Ablatives was charging so, SpaceX being SpaceX, the company decided to build their own heat shield. The only problem with this approach was that SpaceX didn't have the very special- ized facilities required to design and test a heat shield, which is why the company turned to NASA for help. And that is how PICA-X was developed.

SpaceX asked engineers at Ames Research Center (ARC) to help develop Dragon's heat shield and NASA agreed to collaborate [4]. Thanks to the agency's support, the heat shield was designed, developed, and qualified in less than four years. It was an interesting collaborative venture that brought together two very different cultures, which was something of a shock for Dan Rasky. Rasky, who was one of NASA's engineers who had developed the original heat shield material, was sent to work with SpaceX where he worked alongside Andrew Chambers who was SpaceX's project lead for the PICA-X. In one meeting with Elon Musk and an assortment of SpaceX engineers, SpaceX's CEO and CTO asked Rasky what he thought about one of the options for developing PICA.

Rasky replied by outlining his choices and the rationale for those choices. Musk nodded and then announced that SpaceX would follow Rasky's recommendations [5]. For someone used to working in the cumbersome bureaucratic decision-making environment that is NASA, this on-the-spot management style was something of a shock to Rasky. If he had made his proposals to NASA, it would have resulted in a shopping list of studies followed by more meetings that would have inevitably spawned more studies before a decision was made. It was this kind of efficient decision-making that resulted in the rapid four-year in-house development of the PICA-X. Of course, the development and design process was accelerated by the fact that NASA experts and technical support were only an e-mail away. That, combined with the size of the PICA team – just five engineers and a handful of technicians – meant that the whole development process was not only streamlined, but also mutually beneficial because NASA was also learning how some of SpaceX's work practices might be applied to the agency's programs.

Once the design had been agreed upon, SpaceX was able to take advantage of NASA's Atmospheric Re-entry Materials and Structures Evaluation Facility (Arc-Jet) facilities (Figure 3.6) to test the heat shield material at the anticipated re-entry temperatures [6]. Access to the Arc-Jet was critical to SpaceX's development of the heat shield because there was no other way to test the PICA-X material. The PICA-X material stood up to the tests very well, with just a few centimeters needed to withstand temperatures of up to 1,850°C. Why is the material so effective? Well, to answer that, it's necessary to take a closer look at the structure of the material. To begin with, the material isn't the easiest to

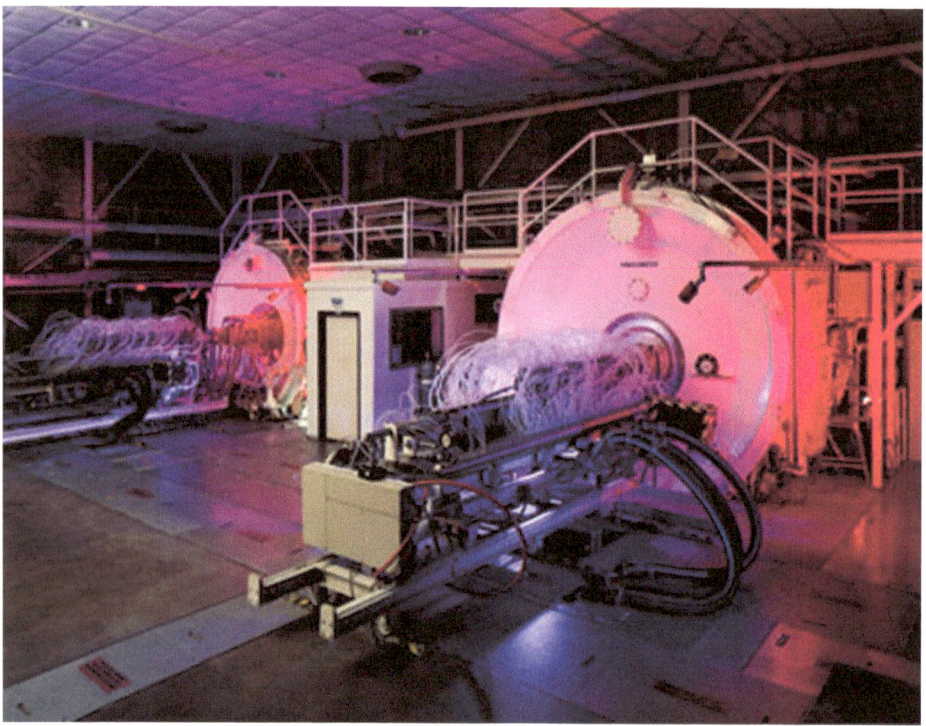

3.6 NASA's Arc-Jet facility where SpaceX developed its PICA-X heat shield. Credit: NASA

apply to the curved structure of a spacecraft because PICA-X isn't that flexible, but what it lacks in flexibility it more than makes up for in heat protection. Because of its lack of flexibility, the material is applied in small sheets to an adhesive substrate. And, because of the material's extremely low conductivity, the layers only need to be applied to a thickness of six centimeters. The PICA-X material also happens to be very, *very* light, with a density of just 0.27 grams per cubic centimeter [7]. The reason such a lightweight material can withstand re-entry temperatures lies in the microstructure of the PICA-X. Remember, the PICA acronym stands for Phenolic Impregnated Carbon Ablator, so let's take a look at these components.

First, let's consider the ablative function of the material. An ablator heat shield does its job through a process of erosion, during which part of the heat shield material pyrolyzes. The PICA heat shield (Figure 3.7) comprises two materials that are identified in its name: phenolic resin and carbon fiber insulator. These two materials are bonded to create the low-density structure of the heat shield and it is the resin that pyrolyzes during re-entry. During re-entry, the very high temperatures cause the resin to sublimate which leaves a carbon matrix and it is this matrix that results in the char [8]. How is the shield manufactured? Well, only SpaceX knows the exact manufacturing process, but we do know that one of the key elements of PICA is Fiberform, which is a carbon fiber insulator. The Fiberform material is treated with phenolic resin and this results in a bonding of the two materials that produces a very temperature-resilient material. Testing of the material revealed that less than a centimeter of material ablated away during a typical re-entry. To reuse the heat shield, technicians simply remove the char and the system is ready to go again.

3.7 SpaceX's Chief Technology Officer inspects Dragon's PICA-X heat shield. Credit: NASA

Reaction control system, power, and recovery

Okay, so that's the heat shield. Now let's get back to the rest of the development and design but, before we do, some may be wondering how Dragon got its name. Well, like all of SpaceX's hardware, there is a story behind this. Dragon was named after the creature in the Peter, Paul and Mary song "Puff the Magic Dragon" for the simple reason that, when Musk began his space venture, there were many who pronounced his goals beyond reach. Incidentally, in case you're wondering, SpaceX's Falcon rocket was named after the Millennium Falcon as piloted by Han Solo in the *Star Wars* series. This tradition of having cool names for its rockets is in marked contrast with NASA, which never seems to get much beyond an acronym. Even the Orion used to be referred to as the Multi-Purpose Crew Vehicle (MPCV). And the Space Station? Well, that's still known as the ISS. You would have thought that, after spending US$100 billion building the thing, the orbiting outpost would have deserved a more catching name, but that's the government for you.

We've talked about the PICA-X heat shield. Now let's move on to the reaction control system (RCS), which Dragon needs to maneuver in orbit and during re-entry. Dragon's RCS (Figure 3.8) comprises 12–18 Draco thrusters, which, like just about everything in the SpaceX inventory, were developed in-house. Each engine, which is fueled by nitrogen tetroxide as oxidizer and monomethyl-hydrazine, produces 400 newtons of thrust. Once in orbit, Dragon uses these thrusters to perform the elaborate and precise attitude control and rendezvous maneuvers when approaching and departing the ISS. And, to ensure those maneuvers are precise, Dragon is fitted with all the necessary navigation equipment: inertial measurement units (IMUs), GPS, iridium recovery beacons, and star trackers. As Dragon makes its approach to the ISS, it adjusts its attitude using the IMUs and star trackers, which enable an accuracy of 0.004° for attitude determination and 0.012° for attitude control on each axis.

And, while Dragon is in orbit, it must regulate temperatures inside the capsule. Temperature control must also be very precise, because some of the cargo bound for the ISS includes biological material and life sciences experiments. To achieve the necessary temperature regulation, Dragon sports two pumped fluid cooling loops and radiators mounted to the trunk's structure. Internally, air is circulated by fans, while sensors ensure temperature is maintained within ±1°C. Depending on the type of cargo, temperature can be adjusted between 10 and 46°C, while humidity can be regulated within 25–75% relative humidity. Dragon is pressurized up to 14.9 pounds per square inch and the pressure can be actively controlled.

Another very important feature while performing those rendezvous maneuvers is communication and telemetry. While Dragon can communicate using NASA's Tracking and Data Relay Satellite (TDRS) system, it can also communicate via ground stations: 300 kbps for command uplink and 300 Mbps for command uplink. Since ferrying cargo to and from the ISS is Dragon's primary role, its payload capacity is obviously especially important. The pressurized volume of Dragon is 10 cubic meters, but that's just the capsule. Another 14 cubic meters of volume are available for external payloads. This volume is found in the trunk (Figure 3.9), which is beneath the capsule. And, if more cargo needs to be flown, SpaceX has a trunk adapter which increases volume to 34 cubic meters.

Dragon's power is provided by two solar arrays (Figure 3.10) that produce up to 2,000 watts and a peak of up to 4,000 watts. The arrays are deployed following insertion into orbit and are jettisoned prior to re-entry: for ascent to orbit and re-entry, Dragon relies on

3.8   Dragon's reaction control system (RCS). Credit: NASA

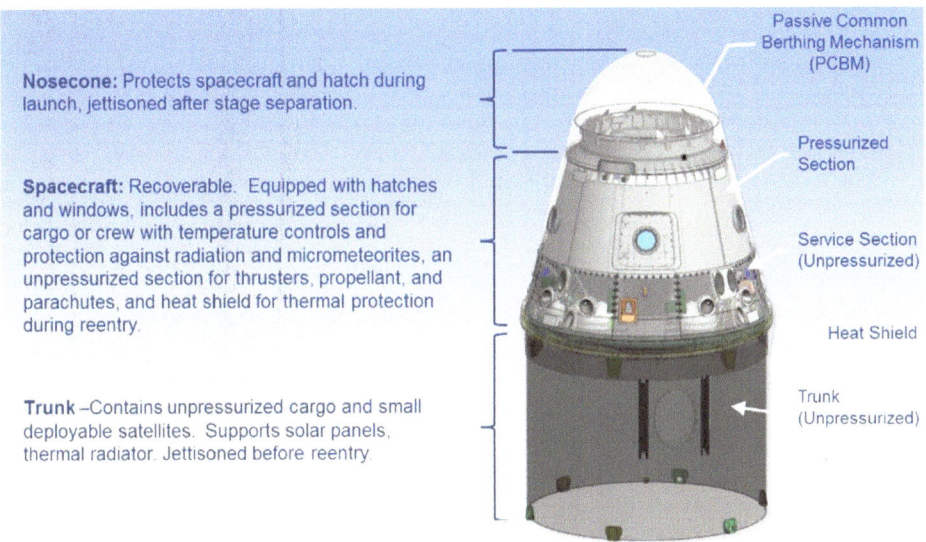

Nosecone: Protects spacecraft and hatch during launch, jettisoned after stage separation.

Spacecraft: Recoverable. Equipped with hatches and windows, includes a pressurized section for cargo or crew with temperature controls and protection against radiation and micrometeorites, an unpressurized section for thrusters, propellant, and parachutes, and heat shield for thermal protection during reentry.

Trunk –Contains unpressurized cargo and small deployable satellites. Supports solar panels, thermal radiator. Jettisoned before reentry.

Passive Common
Berthing Mechanism
(PCBM)

Pressurized
Section

Service Section
(Unpressurized)

Heat Shield

Trunk
(Unpressurized)

3.9   There is plenty of space for cargo in Dragon: up to 6,000 kg divided between the pressure vessel, the trunk, and the sensor bay. Unpressurized payloads are carried in the trunk. Credit: SpaceX

3.10 On its first flight to the International Space Station (ISS), Dragon used its solar arrays for its main power source. The power – 5,000 W – generated by the solar arrays provided sufficient power for heating, cooling, running sensors, and communicating with Mission Control. Why solar power? Earlier spacecraft such as the Shuttle and Apollo used fuel cells but these are limited by how much hydrogen and oxygen can be carried. Batteries also happen to weigh a lot. But solar arrays are compact and they provide reliable and renewable energy. As with so much of SpaceX's technology, Dragon's solar arrays were developed in-house. Credit: NASA

four lithium-polymer batteries. Other elements of Dragon's electrical system are two power buses that provide 120 volts of direct current (VDC) and 28 VDC. Once Dragon has performed all its necessary maneuvers and has been cleared for final approach to the ISS, it needs to make use of its sensors and docking mechanism, which can be either the integral common berthing mechanism (CBM) (Figure 3.11) or the low-impact docking system (LIDS) (Figure 3.12).

Dragon can perform a lifting re-entry to target its ocean landing, which is gentle thanks to dual drogue parachutes that slow and stabilize the craft before three main parachutes deploy (Figure 3.13). The parachute system is fully redundant. Following splashdown, GPS, and iridium locator beacons can be used to locate the vehicle if necessary. Recovery is performed via ship and the capsule enters its refurbishment processing.

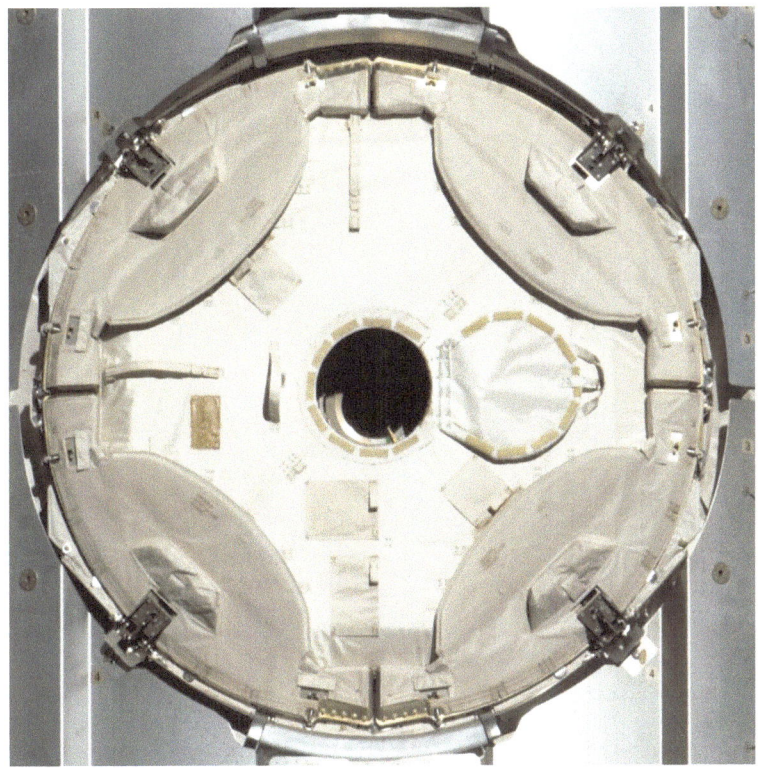

3.11   The common berthing mechanism (CBM). Credit: NASA

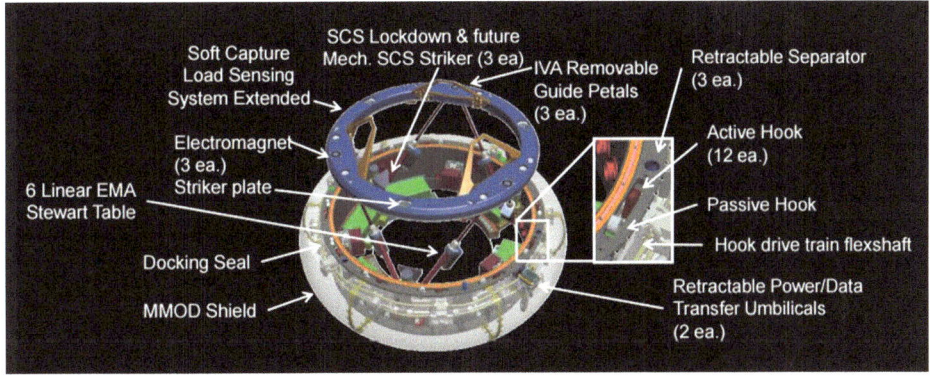

3.12   NASA's low-impact docking system (LIDS). Credit: NASA

3.13 Dragon being recovered after its maiden flight on 8 December 2010. Credit: SpaceX

DEBUT DRAGON: FLIGHT #1

At 14:45 Eastern Daylight Time (EDT) on 4 June 2010, a prototype Dragon spacecraft was launched atop SpaceX's Falcon 9 launch vehicle. Launch, which came at the very tail end of a four-hour launch window, had been interrupted by incursions into the range, issues with the Flight Termination System (FTS), and a last-minute abort. For some observers, the delays were an uncomfortable reminder of the problems encountered when launching the Falcon family of rockets: the first five flights of the smaller Falcon 1 resulted in three failures and two successes. Preparations for the historic flight began when the Falcon 9 and the Dragon Space Qualification Unit (SQU) were rolled onto the launch pad at Cape Canaveral in February 2010. One of the first tests to be conducted was a static firing, but the first attempt was delayed due to problems with a ground isolation valve. Four days later, the test was completed successfully, clearing the way for launch, but this was delayed several times due to problems with the FTS, which is used by the Range Safety Officer in the event that a rocket goes off course (this happened in August 1998, when a Delta III was destroyed). Finally, after much trouble-shooting, 4 June 2010 rolled around and the countdown began. SpaceX engineers and Elon Musk crossed their fingers. There was an awful lot riding on this, not just for commercial spaceflight, but for manned spaceflight. A successful launch would go some way to validating President Obama's plan to reconfigure NASA by using commercial companies to ferry astronauts to the ISS. A failure? Well, that didn't bear thinking about, and that's why there were so many nervous people at the launch site that day. After all, the Falcon 1 hadn't had a sterling launch record, and a failure of the Falcon 9 would be interpreted by some as a red flag. People would call into question this commercial spaceflight lark and would start wondering whether this business of letting private firms launch astronauts into orbit might not be such a good idea after all. Which was why Elon Musk had cautioned against expectations that were too high:

> "100 percent success would be reaching orbit. Given that this is a test flight, whatever percentage of getting to orbit we achieve would still be considered a good day. If just the first stage functions correctly, it's a good day. It's a great day if both stages function."
>
> *SpaceX CEO, Elon Musk*

Despite the words of caution, there was no denying that this launch was pivotal, not only for Musk's integrity, SpaceX's credibility, but also the President's long-term vision for how America's manned space program might pan out over the next few years. Some portrayed the Falcon 9 flight as hanging the future of commercial spaceflight on one launch, although that was pushing it a little. In 2010, New Space had developed a broader base, but there was still the sense that a failure would turn the clock back significantly. Musk? He reckoned his first Falcon 9 had about a 70–80% chance of success. Those were pretty good odds in a business as volatile as the commercial launch business. And, in many ways, just getting the Falcon 9 to the launch pad was a success, given that SpaceX had spent just 10% of what NASA had spent preparing its Ares launcher.

The clock started ticking on SpaceX's launch preparations at 10:56 EDT on 4 June 2010. The company had a four-hour launch window and weather forecasters predicted a

60% chance of the weather cooperating. At 11:23 EDT, range control asked SpaceX to hold the count at T–15 minutes to troubleshoot a telemetry connection. Half an hour later, SpaceX was still waiting for a new T–0 time. Then, at 12:22, a blocked signal required more trouble-shooting and, as if that wasn't enough, at 12:56, a boat wandered too close to the launch pad. SpaceX recalibrated for a 1:30 EDT launch to allow time for the boat to get its bearings, and the count resumed. Just before launch was due, a last-minute glitch forced a last-minute abort at 1:33 EDT. Once again, SpaceX reworked the plan and announced it would recycle the count. The launch was still on. Following the reset, a launch time of 2:45 was announced and, finally, at 2:46, Falcon 9 lifted off on its maiden flight (Figure 3.14). Nine minutes later, the Falcon 9's second-stage engines shut down and the prototype Dragon (see sidebar) had reached orbit (Tables 3.1 and 3.2).

3.14 Falcon 9 carrying the Dragon Space Qualification Unit (SQU) launches from Launch Complex 40 on 4 June 2010. Credit: SpaceX

Table 3.1 Dragon specifications.

Length	2.9 m
Diameter	3.6 m
Sidewall angles	15°
Pressurized volume	10 m³
Unpressurized volume	14 m³
Trunk extension	34 m³
Sensor bay	0.1 m³
Mass	4,200 kg
Launch payload	6,000 kg
Return payload	3,000 kg
Endurance	Up to 2 years
Maximum crew	7
Avionics	Full redundancy
Reaction control	18 Draco thrusters
Propellant	Hydrazine/nitrogen tetroxide
Propellant mass	1,290 kg
Docking mechanism	LIDS or APAS
Power supply	2 solar arrays: 1,500–2,000 W
Power buses	28 and 120 VDC
Batteries	4 Li-polymer batteries
Cabin pressure	13.9–14.9 psi
Cabin temperature	10–46°C
Cabin humidity	25–75%
Command uplink	300 kbps
Downlink	>300 Mbps
Windows	Up to 4
Window diameter	30 cm

Table 3.2 Falcon 9/Dragon SQU flight summary.

Mission type	Demonstration	*Rocket*	Falcon 9 v1.0 F1
COSPAR ID	2010-026 A	*Launch site*	Cape Canaveral SLC-40
SATCAT no.	36595	*Decay date*	27 June 2010
Orbits completed	359	*Perigee*	249.5 km
Launch mass	333,400 kg	*Apogee*	252.5 km
Launch date	4 June 2010	*Inclination*	34.5°

Dragon Spacecraft Qualification Unit (SQU)

The Dragon SQU had been used as a ground-test article at SpaceX's Hawthorne, California, location. The spacecraft comprised a nose-cone, which protected the vessel during ascent; the spacecraft itself, which included a service compartment; and the trunk, which was designated for stowage of unpressurized cargo.

After inspecting the boilerplate article to assess shape, mass, and various other tests, the SQU orbited Earth more than 300 times before re-entering the atmosphere on 27 June 2010.

COTS DEMONSTRATION FLIGHT #1: FLIGHT #2, 8 DECEMBER 2010

"There's so much that can go wrong and it all went right. I'm sort of in semi-shock."
Elon Musk speaking at a press conference
following the successful C1 Mission

The objective of the maiden flight of the Dragon had been primarily to relay aerodynamic data during the ascent because the capsule wasn't designed to survive re-entry. The C1 flight was a little more ambitious because SpaceX planned retrieving the capsule, but first they had to wait for the Federal Aviation Administration (FAA) to issue a re-entry license, which is what the agency did on 22 November 2010 – the first such license ever awarded to a commercial vehicle. Three weeks later, Dragon launched (Figure 3.15) on COTS Demo Flight 1.

"When Dragon returns, whether on this mission or a future one, it will herald the dawn of an incredibly exciting new era in space travel. This will be the first new American human capable spacecraft to travel to orbit and back since the Space Shuttle took flight three decades ago. The success of the NASA COTS/CRS program shows that it is possible to return to the fast pace of progress that took place during the Apollo era, but using only a tiny fraction of the resources. If COTS/CRS continues to achieve the milestones that many considered impossible, thanks in large part to the skill of the program management team at NASA, it should be recognized as one of the most effective public-private partnerships in history."

Elon Musk

3.15 Falcon 9 carries Dragon on its maiden mission from Launch Complex 40 at Cape Canaveral on 8 December 2010. Credit: NASA/Alan Ault

CUCU: DRAGON'S COMMUNICATION UNIT

While SpaceX was busy preparing for the SQU flight, NASA was making its own preparations on board the ISS. These preparations dealt with the Commercial UHF Communication Unit (CUCU), which was a key item of hardware that was needed for upcoming Dragon flights to the orbiting outpost. The CUCU had been flown to the ISS in 2009 on board the Shuttle *Atlantis* during STS-129.

The CUCU was ferried up to the ISS so that crewmembers could use it to monitor approaching and departing Dragon's once cargo delivery missions started. One of the demonstration tests that utilized the CUCU was Dragon's ability to establish communication and relative GPS with the orbiting outpost at a range of 23 kilometers before performing a fly-by at a distance of 10 kilometers below the ISS. Assuming these tests went well, the Dragon would perform one final demonstration, which would require the capsule to rendezvous with the ISS and be captured by the station's remote manipulator system, before being mated with the Node 2 nadir port. Once the CUCU arrived on board the ISS, Jeff Williams worked with SpaceX engineers and NASA Mission Control to perform the necessary checkout procedures. These procedures continued through March 2010, when SpaceX and NASA tested the system to send signals between the orbiting station and NASA's Dryden ground station to establish a baseline of radio frequencies for mission operations.

DRAGONEYE

In addition to the CUCU, NASA was also preparing another item of equipment required to test Dragon's ability to rendezvous with the ISS. Dubbed DragonEye, this proximity device was a Laser Imaging Detection and Ranging sensor (LIDAR) that was designed to provide three-dimensional images of range and bearing information from the Dragon to the station. It was tested successfully during the Shuttle *Endeavour*'s STS-127 flight, where the DragonEye was mounted on the Trajectory Control System carrier assembly on the Shuttle's docking system.

SpaceX targeted 7 December 2010 as the launch date for COTS 1. In preparation for the launch, the company planned to conduct static firing tests, the first of which, on 3 December, was aborted at T–1.1 seconds as a result of high engine-chamber pressure. The following day, the static fire resulted in another abort at T–1.9 seconds as a result of low gas generator pressure on engine #6. The countdown clock was recycled and SpaceX made a third attempt later that day that resulted in a full-duration fire. All was set for launch on 7 December but, the day before launch, engineers discovered two small cracks in the second-stage Merlin engine nozzle extension (this part of the engine increases engine efficiency in a vacuum), which prompted a systems check to make sure it wasn't part of a bigger problem. Finally, on 8 December, SpaceX was ready to launch.

Dragon C1 planned mission events (sourced and adapted from the SpaceX Press Kit)

Countdown

T–02:35:00	Chief Engineer polls stations. Countdown master auto-sequence proceeds with Liquid Oxygen (LOX) load, RP-1 fuel load, and vehicle release
T–01:40:00	Allow countdown master auto-sequence to proceed into lowering the strong-back
T–00:60:00	Allow the master auto-sequence to proceed with stage 2 fuel bleed, stage 2 thrust vector control bleed. Verify all sub-auto-sequences in the countdown master auto-sequence have been performed, except for terminal count
T–00:13:00	SpaceX Launch Director polls readiness for launch
T–00:11:00	Logical hold point if launch point

Terminal count (begins at T–10 minutes)

T–00:09:43	Open pre-valves to the nine stage 1 engines and begin chilling Merlin engine pumps
T–00:06:17	Command flight computer to enter alignment state
T–00:05:00	Stop loading of GN2 into ACS bottle on stage 2
T–00:04:46	Transfer to internal power on stage 1 and stage 2
T–00:03:11	Begin arming FTS
T–00:03:02	Terminate LOX propellant topping, cycle fuel trim valves
T–00:03:00	Verify movement on stage 2 thrust vector control actuators
T–00:02:30	SpaceX Launch Director verifies GO
T–00:02:00	Range Control Officer verifies range is GO
T–00:01:35	Terminate helium loading
T–00:01:00	Command flight computer state to start up
T–00:01:00	Turn on pad deck and Niagara Water
T–00:00:50	Flight computer commands thrust vector control actuator checks on stage 1
T–00:00:40	Pressurize stage 1 and stage 2 propellant tanks
T–00:00:03	Engine controller commands engine ignition sequence to start
T–00:00:00	*Lift-off*
T+0:02:58	Stage 1 shutdown (main engine cut-off)
T+0:03:02	Stage 1 separates
T+0:03:09	Stage 2 engine start
T+0:09:00	Stage 2 engine cut-off
T+0:09:35	Dragon separates from Falcon 9 and initializes propulsion
T+0:13	On-orbit operations
T+2:32	De-orbit burn begins
T+2:38	De-orbit burn ends
T+2:58	Re-entry phase begins (entry interface)
T+3:09	Drogue chute deploys
T+3:10	Main chute deploys
T+3:19	Water landing

Dragon C1 actual mission events[3]

Countdown

T–00:20:00. 07:46 CST	Range cleared for launch
T–00:14:00. 07:54 CST	Terminal countdown initiated
T–00:07:00. 08:03 CST	Countdown aborted
08:11 CST	SpaceX controllers review data. Countdown clock reset to T–13 minutes
08:13 CST	New launch time of 09:42 CST announced
08:39 CST	New launch time of 09:43 CST announced
T–00:13:00. 09:30 CST	Countdown clock restarts
T–00:10:00. 09:33 CST	Engine pre-valves open
T–00:07:00. 09:36 CST	Command flight computer entered alignment. Stage 1 and stage 2 transferred to internal power at T–4 minutes 46 seconds
T–00:05:00. 09:38 CST	Falcon 9 on internal power
T–00:04:00. 09:39 CST	FTS armed
T–00:03:00. 09:40 CST	SpaceX Launch Director confirms GO for launch. Range Officer confirms GO for launch
T–00:02:00. 09:41 CST	Thrust vector control actuator checks on stage 1
T–00:01:00. 09:42 CST	Tanks pressurizing

Lift-off at 09:43 CST

T+00:02:00. 09:45 CST	Stage 1 shutdown
T+00:03:00. 09:46 CST	Stage separation
T+00:04:00. 09:47 CST	Stage 2 fired
T+00:05:00. 09:48 CST	Dragon separates from Falcon 9
T+00:09:00. 09:52 CST	On-orbit operations commence
T+00:11:00. 09:54 CST	Dragon in orbit

Orbital operations

T+00:59:00. 10:48 CST	Dragon operating as planned
T+01:27:00. 11:16 CST	Dragon communicates with TDRSS
T+01:28:00. 11:17 CST	Stage 1 recovered
T+02:28:00. 12:17 CST	Draco thrusters performed de-orbit burn
T+02:33:00. 12:22 CST	De-orbit burn completed
T+02:34:00. 12:23 CST	Dragon in re-entry attitude

Recovery

T+03:03:00. 12:52 CST	Drogue chutes deployed
T+03:06:00. 12:55 CST	Main parachutes deployed
T+03:15:00. 13:04 CST	Splashdown
T+03:38:00. 13:27 CST	Floats attached to Dragon

---

[3] Sourced from blog by Robert Pearlman, Editor, *CollectSpace*.

Table 3.3 Falcon 9/Dragon C1 flight summary.

Rocket	Falcon 9 v1.0
Launch site	Cape Canaveral SLC-40
Perigee	288 km
Apogee	301 km
Inclination	34.53°
Landing site	Pacific Ocean, 800 km west of Baja, Mexico

All in all, Dragon's demonstration flight was very successful – and historic, because Dragon's orbital journey marked the first time a commercial company had ever recovered a spacecraft that had re-entered from orbit (Table 3.3). This also placed SpaceX alongside the six countries and government agencies that had previously achieved the feat: Russia, the US, China, Japan, the European Space Agency, and India.

Post flight

"It's just mind-blowingly awesome. I wish I was more articulate, but it's hard to be articulate when you're mind's blown, but in a very good way."

Elon Musk, following the Dragon's C1 mission

"After today, we have increased confidence in SpaceX systems, launch vehicles and spacecraft."

Alan Lindenmoyer, head of COTS program

"This is the first in a new generation of commercial launch systems that will help provide vital support to the International Space Station and may one day carry astronauts into orbit. This successful demonstration flight is an important milestone in meeting the objectives outlined by President Obama and Congress, and shows how government and industry can leverage expertise and resources to foster a new and vibrant space economy."

NASA Administrator Charles Bolden

Before Dragon's C1 flight, Musk had predicted a 60% chance of success, so he had every reason to be ecstatic after such a successful mission which, according to Alan Lindenmoyer, head of the COTS program, checked all the mission objectives. At the time of Dragon's flight, COTS had invested US$253 million in SpaceX and it appeared that it had been money well spent. The success of the C1 mission also opened the door to accelerate the test strategy by combining flights C2 and C3. Prior to C1, the plan had been to fly the C2 mission to demonstrate Dragon's ability to approach the ISS before returning to Earth. The C3 mission would then demonstrate docking with the orbiting outpost once Dragon had been grappled by the station's robotic arm.

A series of internal meetings discussing the feasibility of SpaceX's C2/C3 mission were held at Johnson Space Center during the early months of 2011. Mission managers had to consider what work was left to be done, launch date estimates, and the impact on manifests. As of February 2011, the ISS long-range manifest still listed C2 and C3 as separate missions, with C2 due to launch on 15 July 2011 followed by C3 on 8 October 2011. Following the original manifest would have resulted in the first operational Dragon mission – CRS-1 – being launched in mid-December 2011, which was shaping up to be a very busy time for the ISS, with planned visits by the Soyuz and Progress in addition to Orbital's Cygnus demonstration flight.

In March 2011, a memo listing the additional milestones for the C2/C3 mission was submitted to Doug Cooke, NASA Associate Administrator, before a Joint Program Review with the station's management took place. This meeting was followed by a Dragon C3 rendezvous simulation on 23 March which in turn was followed by more memos in April that suggested NASA was on board for a combined mission. The Flight Operations Review (FOR) then suggested an August launch for the C2/C3 mission but a Technical Interchange Meeting (TIM) for Dragon Operations conducted at SpaceX's Hawthorne site decided on a launch date of 30 November 2011. Using SimCity, SpaceX continued modeling the challenges of the C2/C3 mission which included designing a mission profile that would permit Dragon to combine the C2 flight objectives (performing a flight within 10 kilometers of the ISS) with those of the C3 mission (approach, capture, and docking). A decision point was scheduled for June, following numerous Dragon Special Topics meetings and meetings with the Commercial Cargo department. SpaceX received the go-ahead for its C2/C3 mission in July 2011 for a 30 November 2011 launch date.

REFERENCES

1. Grantham, K. DragonLab Data Sheet [online article], Space-X, *www.spacex.com/downloads/dragonlabdatasheet.pdf* (2012).
2. Mars Science Laboratory Heat Shield. Ames Research [online blog], *www.nasa.gov/centers/ames/research/msl_heatshield.html* (2012).
3. Marra, F.; Fossati, F.; et al. Carbon–Phenolic Ablative Materials for Re-Entry Space Vehicles: Manufacturing and Properties. Composites: Part A [online article], Elsevier, *www.sciencedirect.com/science/article/pii/S1359 835X10001776* (2010).
4. Stackpoole, M.; Thornton, J.; Fan, W. Ongoing TPS Development at NASA Ames Research Center [online article], Ames Research Center, *http://ntrs.nasa.gov/archive/nasa/casi.ntrs.nasa.gov/201 10013352_2011014017.pdf* (2012).
5. Chambers, A.; Rasky, D. NASA SpaceX Work Together [online article]. *Ask Magazine*, **40**, *www.nasa.gov/pdf/489058main_ASK_40_space_ x.pdf* (2010).
6. Chambers, D. Arc-Jet Testing Facility [online article], Johnson Space Center Engineering, *www.nasa.gov/centers/johnson/engineering/human_space_vehicle_systems/atmospheric_reentry_materials/index.html* (January 22, 2013).
7. Walker, R. Density of Materials SI Metric [online article], *www.simetric.co.uk/si_materials.htm* (April 4, 2011).
8. Shanklin, E. SpaceX Manufactured Heat Shield Material Passes High Temperature Tests Simulating Re-entry Heating Conditions of Dragon Spacecraft [online article], Space-X, *www.spacex.com/press.php?page=20090223* (2013).

4

Dragon at the International Space Station

Roll-out of the Falcon 9. Credit: SpaceX

© Springer International Publishing Switzerland 2016

E. Seedhouse, *SpaceX's Dragon: America's Next Generation Spacecraft*,
Springer Praxis Books, DOI 10.1007/978-3-319-21515-0_4

With the C2/C3 flight approved, SpaceX and NASA knuckled down to the planning ahead of the November flight, but it was to prove a tortuous journey. The first black cloud on the horizon was dealing with the Russians who weren't too enthused about allowing a commercial vehicle to dock with the International Space Station (ISS). A post on RIA Novosti's website stated:

> "'The U.S. private space capsule Dragon will conduct a flight near the International Space Station, but docking between them is not planned,' Vladimir Solovyov, head of the Russian segment of the ISS mission control center said on Friday. 'We will not issue docking permission unless the necessary level of reliability and safety is proven,' said Alexei Krasov, head of the human spaceflight department of Roscosmos. 'So far we have no proof that this spacecraft duly comply with the accepted norms of spaceflight safety.'"

So much for collaboration! Worse was to come on 24 August 2011 when a Russian Progress cargo ship carrying tonnes of supplies for the ISS crashed less than six minutes into flight. This was a big problem because the Progress and the Soyuz use the same booster and an investigation into the accident would delay the arrival of the crew trained to berth the Dragon. Initially, a revised launch date of 19 December 2011 was announced, but this was later updated to January 2012.

Two months after the Progress accident, the news was a little more upbeat with the arrival of Dragon at Launch Complex 40 (Figure 4.1) after having been trucked from SpaceX's Hawthorne facility. Prior to its cross-country trip, Dragon had been subjected to a thermal vacuum test, which it had passed with flying colors. The arrival of Dragon at Cape Canaveral occurred shortly after SpaceX had passed a Preliminary Draft Review (PDR) of its capsule's Launch Abort System (LAS). Dragon's reusable LAS was unlike conventional single-use systems because DragonRider – SpaceX's name for its LAS – was built into the spacecraft's hull whereas abort systems comprise a small rocket that sits on top of the spacecraft.

4.1   CRS2/3/CRS 2+ on the launch pad. Credit: NASA

Shortly after arriving at SLC-40, SpaceX technicians and engineers began to prepare Dragon for launch. This required testing the new features such as the hatch, the vehicle's grapple fixture, and a claw that provided data connections between Dragon and the trunk. With the Falcon 9 booster already on site, the only element SpaceX was waiting for was the service module/trunk, which was due the following week. Preparations continued on schedule, but there was to be no December flight. On 5 December 2011, SpaceX announced the flight would now probably be flown in the February–March time frame. One of the reasons for the delay was the Russians, who weren't happy with the software updates supplied by SpaceX, although Roscosmos and RSC Energia had signed approval for Dragon's arrival at the ISS. One of the misgivings highlighted by the Russians was Dragon's approach and arrival point: the Russians preferred that Dragon use the same approach points as the European Automated Transfer Vehicle (ATV) and Japanese H-I Transfer Vehicle (HTV) used on their maiden flights. While this bureaucratic tug-of-war was taking place on Earth, crewmembers on board the ISS were preparing for Dragon's arrival by digging out the Commercial UHF Communications Unit (CUCU) that was needed during Dragon's approach and rendezvous. At this time, Dragon couldn't be launched anyway because there was only one crewmember – Dan Burbank – who had completed the necessary ground training on the unit. Mission rules required that there be two crewmembers trained on the unit and Don Pettit, the other crewmember with that training, wasn't due up until the Progress issue had been resolved.

In addition to the Russians' concerns, NASA mission managers had their own worries when SpaceX decided to add two ORBCOMM satellites to the Falcon 9 carrying Dragon. This caused some concern over the risk of collision with the ISS, prompting NASA to conduct Monte Carlo analysis to troubleshoot the issue. Then, on 9 December 2011, NASA announced a 7 February 2012 launch date for the C2/C3 flight.

> "There is still a significant amount of critical work to be completed before launch, but the teams have a sound plan to complete it and are prepared for unexpected challenges. As with all launches, we will adjust the launch date as needed to gain sufficient understanding of test and analysis results to ensure safety and mission success."
>
> *William Gerstenmaier*

The reason for the delay was for more safety checks to be completed. NASA didn't elaborate on the nature of the safety checks and didn't specify whether the checks were for Dragon or for the launch vehicle. What was known was that the mission profile had passed the ISS Post-Qualification Review, a significant step that cleared the way for SpaceX to complete the final steps prior to launch. One of those steps was a Wet Dress Rehearsal (WDR) for the Falcon 9, but this was also delayed following the announcement of the postponement to the launch. More bad news followed shortly afterwards when it was announced the launch had been postponed to late March. At this time, SpaceX was under an awful lot of pressure juggling myriad projects, including the commercial orbital transportation system (COTS), the Falcon 9 manifest, developing the Falcon Heavy, Grasshopper, Strato-launch, and the growing backlog of payloads. While the image the company portrays is one of an organization that is cool and calm under pressure, the reality was anything but. One of my colleagues works for SpaceX and he (I can't name him or he'd lose his job, or be sued) alluded to the fact that Elon Musk could be a very difficult boss to work for, partly because Musk seems to occupy a distorted version of reality in which there is no life beyond SpaceX. He also – according to my colleague – is a harsh taskmaster and doesn't believe a job that should take six months can't be done in a week. So, when the March launch date slipped into April, tensions at SpaceX were ratcheted up another notch. The latest postponement was announced alongside the news that the launch of the next expedition crew had also been delayed due to ongoing problems with the Soyuz.

As March rolled around, SpaceX continued their launch preparations by conducting the postponed WDR at Cape Canaveral. The WDR was a significant Launch Readiness Test (LRT) that served as a full dress rehearsal of the countdown. At the T−100 minutes mark of the countdown, the strong-back used to support the Falcon 9 on the pad was lowered, and fuel and vector control was performed at the T−60 minutes mark. A terminal count commenced 10 minutes before launch and the Falcon 9 was transferred to internal power less than five minutes before launch. Countdown events continued all the way down to the T−5 seconds mark, thereby successfully accomplishing the WDR tasks.

Two weeks after the WDR, yet another launch date was announced – this one for 30 April 2012. Following the WDR, one of the next key tests to be performed was the Crew Interface Test (Figure 4.2), which was conducted on 28 March. Working together with SpaceX flight controllers, astronaut Megan McArthur (a member of the STS-125 crew) spent time familiarizing herself with Dragon, testing compatibility of the equipment and systems on board the capsule with the procedures used by NASA flight controllers and crewmembers on board the ISS.

4.2   NASA astronaut Megan McArthur conducting a crew equipment interface test at Space Launch Complex 40 prior to the launch of CRS2/3. The test, which took about five hours, is a procedure that began in the Shuttle era and is a feature of all SpaceX prelaunch procedures. Credit: Paul Bonness/SpaceX

Two days after the Crew Interface Test, technicians began stowing the cargo due for the orbiting outpost: all 530 kilograms of it (see Appendix I for the complete manifest). Much of the cargo comprised water, food, and clothing in addition to an assortment of research payloads and equipment to support the NanoRacks experiments. As Dragon was being loaded, NASA announced that its Flight Readiness Review (FRR) had cleared Dragon for visiting the ISS. The planned mission required Dragon to perform far-field phasing maneuvers on Flight Day 2 (FD2) as depicted in Figures 4.3, 4.4, 4.5, and 4.6.

While performing orbital aerobatics, Dragon's GPS would be tested and free drift and abort capabilities would be assessed. If all these tests were green-flagged, Dragon would be given the go-ahead for a 2.5-kilometer fly-under of the ISS to test UHF communication. Assuming all the FD3 tests were completed successfully, mission managers of the ISS Mission Management Team (IMMT) would approve the objectives of FD4. FD4 would feature Dragon performing a rendezvous with the ISS – well almost, since the Dragon wouldn't approach closer than 30 meters from the station, where it would be grappled by the robotic arm before being berthed to the Node 2 Nadir port. The maneuver is known as a "free-flyer capture" and Dragon wouldn't be the first spacecraft to have been grappled: the Japanese HTV-1 and HTV-2 missions also performed the same maneuver. Once Dragon had been mated with Node 2 using the common berthing mechanism (CBM),

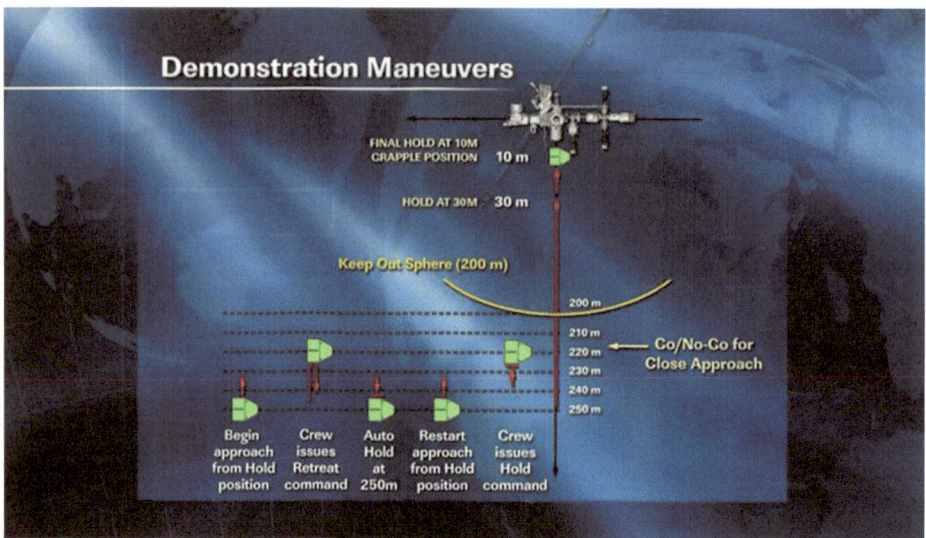

4.3   Demonstration maneuvers. Credit: NASA

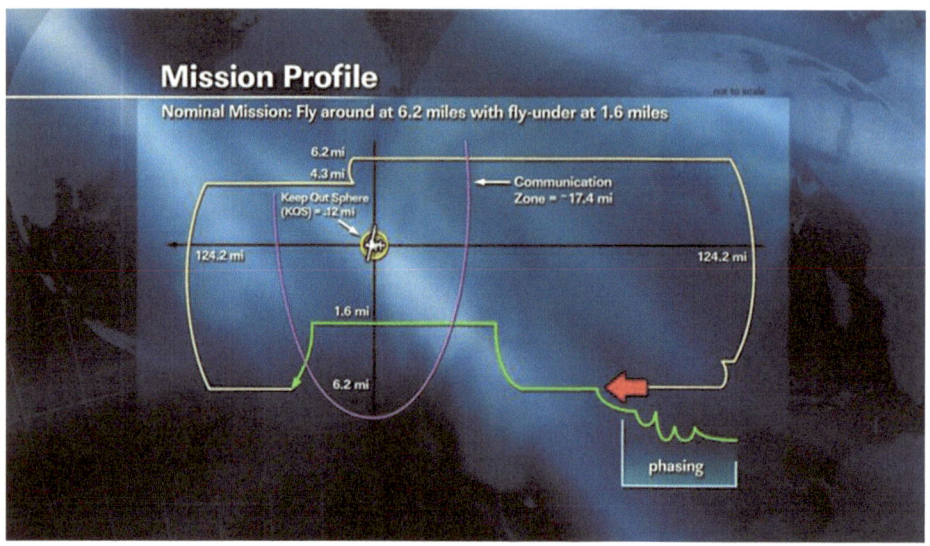

4.4   Mission profile. Credit: NASA

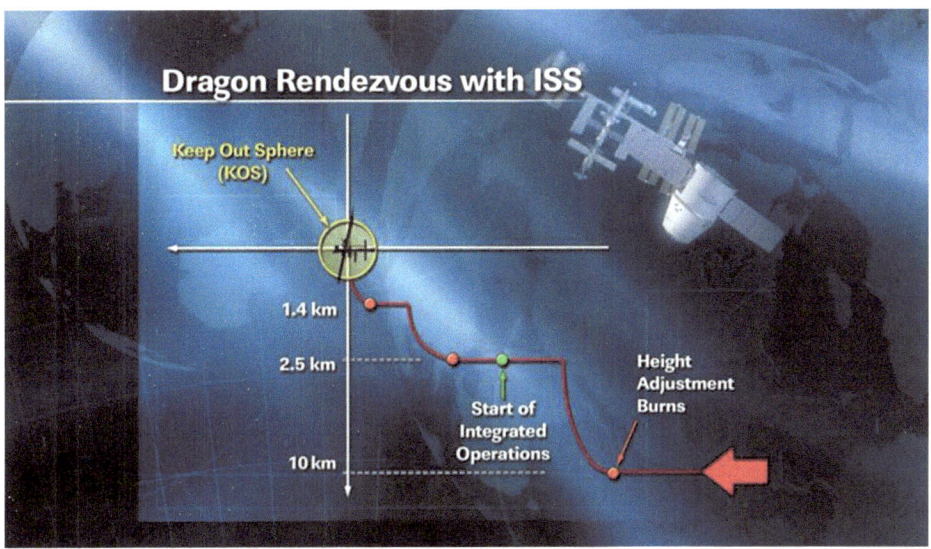

4.5 Rendezvous maneuvers. Credit: NASA

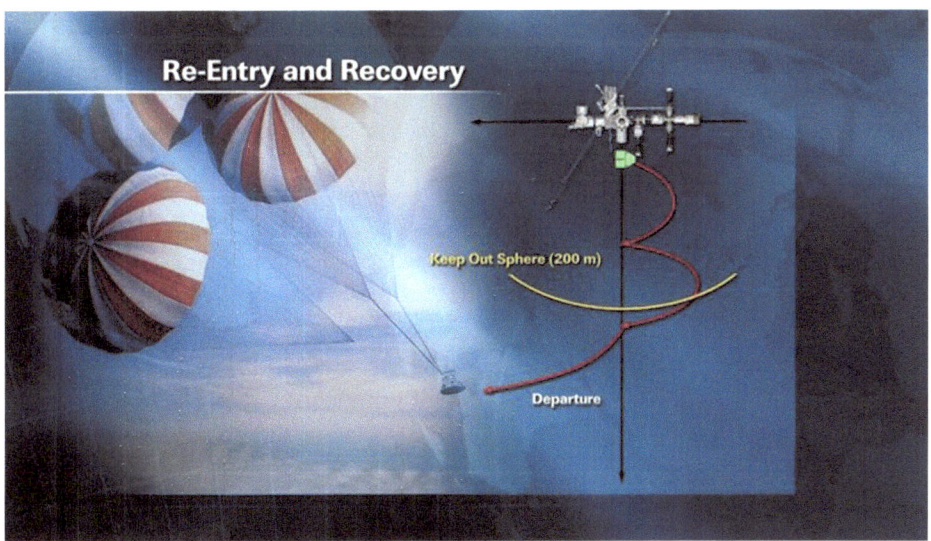

4.6 Re-entry and recovery. Credit: NASA

4.7   Karen Nyberg takes in the view from the Cupola. Credit: NASA

power cables and ventilation hoses would be aligned on FD5, after which Dragon would stay for at least two weeks. At the end of Dragon's visit, the capsule would be unberthed and ungrappled by the robotic arm ready for re-entry and splashdown.

Five days after the Crew Interface Test, the Stage Operations Readiness Review (SORR) was completed. The SORR, which was performed on 5 April 2012, went well, but it did highlight a concern over the Cupola. The Cupola (Figure 4.7) is the observation module that would be used by the crew during the robotic arm manipulation of Dragon. Comprising seven windows, the module provides the crew with great situational awareness when conducting robotic arm operations, but there was a problem with Window #6. The problem was a temperature divergence between two of the sensors, one of which indicated an upper temperature that exceeded 37.3°C for 13 minutes. This was a problem because, under flight rules, any window that exceeded 36.7°C for more than five minutes had to be shuttered. After some discussion with mission managers, the ISS crew attempted a troubleshooting exercise by closing the shutter for Window #1 to try and drop the temperature. Meanwhile, engineers tried to come up with a reason for the temperature spike and suggested that either the seal around the window had been compromised or there was partial delamination of the window, as noted in the SORR:

"On 25/03/12, window scratch panes (with heaters) were R&R for ATV-3 antenna photography. Leading theory is partial delamination of RTD 2 is possible; may have exacerbated during scratch pane removal (not easily accessible for inspection)."

In addition to the window issue, the SORR also noted a problem with the GPS equipment that Dragon would use to communicate with the ISS. The equipment in question was the Space Integrated GPS/Inertial Navigation System, or SIGI. There are two SIGIs on board the ISS and one of them had shown signs of wear and tear. While two SIGIs were required by SpaceX's Launch Commit Criteria (LCC), only one was needed for the rendezvous portion of the flight. The solution was that, if rendezvous data were not available from one of the SIGIs, Dragon would abort its rendezvous. With the SIGI issue resolved, mission managers turned their attention to yet another issue, this being with one of the petals of the Node 2 Active Common Berthing Mechanism (ACBM). The ACBM has four petals, which serve to protect the docking mechanism when a spacecraft is closing in for a hard dock.

The problem in this case was that one of the petals was slightly open. This wasn't a big problem but, if the petal was to become stuck in the open position, the ACBM ring would be exposed to Micro Meteoroid Orbital Debris (MMOD) strike. Equally, if the petal was to become stuck in the closed position, mission managers would have to call off the rendezvous. After viewing the offending petal from all sorts of camera angles and putting it through several successful open and close cycles, mission managers agreed that the petal was not warped. The mission was still on, although it wasn't long before NASA managers announced yet another slip to the launch date: on 23 April, it was announced that the new date would now be 7 May 2012, with a backup date of 10 May. The reason? It had been decided that more time was needed to perform hardware-in-the-loop system testing or, more specifically, Dragon Force testing. For the C2/C3 mission (now renamed C2+) to be successful, there was an awful lot riding on software performing perfectly and one of the simulations SpaceX had yet to run was a hardware-in-the-loop system (Dragon Force) which permits several hardware and software scenarios to be run simultaneously.

While SpaceX conducted its Dragon Force simulation, NASA mission managers struggled to find more launch slots in case the 7 May opportunity didn't work. May was turning out to be a very busy time for ISS traffic, with Soyuz flights arriving and departing, Progress flights coming and going, and also ATV flights. With so much activity, mission managers had to work to de-conflict traffic, because they didn't want a Soyuz vehicle and Dragon arriving at the same time. With the 7 May Falcon 9/Dragon launch date, there was little room for slippage because the Russians were planning a 15 May launch of the Soyuz. So, if there was a delay launching Dragon, the likely backup date would probably have to be 19 May to avoid having two spacecraft on top of one another.

On 30 April, Space X conducted a static fire test of the Falcon 9 rocket. Even this wasn't without issues as a computer gremlin resulted in an abort of the first attempt, but the delay was brief and the launcher's engines were ignited for two seconds. For a while, it looked as if the 7 May date was a go. But it wasn't. NASA mission managers had decided to postpone the 7 May and 10 May launch dates until after the Soyuz had docked because, if some glitch prevented launch, there would be no backup opportunity. So the launch date was reworked for a 19 May launch (Tables 4.1a, 4.1b, and 4.1c) with a backup launch date of 22 May.

On May 19, the stage was set for the C2+ launch. Blast-off was scheduled for 4:55 Eastern Daylight Time (EDT) but, shortly after the ignition sequence started, a high pressure reading was detected in Engine #5, prompting a shutdown and launch abort.

Table 4.1a Flight Day 1 prelaunch events.

Time to lift-off, H:M:S	Event
7:30:30	Vehicles are powered on
3:50:00	Commence loading liquid oxygen (LOX)
3:40:00	Commence loading RP-1 (rocket-grade kerosene)
3:15:00	LOX and RP-1 loading complete
0:10:30	Falcon 9 terminal count auto-sequence starts
0:05:30	Dragon terminal count auto-starts
0:02:30	SpaceX Launch Director verifies GO for launch
0:02:00	Range Control Officer (USAF) verifies range is GO for launch
0:01:00	Command flight computer state to start-up, turn on pad deck and Niagara Water
0:00:40	Pressurize propellant tanks
0:00:03	Engine controller commands engine ignition sequence to start
0:00:00	Falcon 9 launch

Source: International Space Station partnership.

Table 4.1b Flight Day 1 launch events.

Time after lift-off, H:M:S	Event
0:01:24	Max Q
0:03:00	Stage 1 engine shutdown/main engine cut-off (MECO)
0:03:05	Stages 1 and 2 separate
0:03:12	Stage 2 engine starts
0:03:52	Dragon nose-cone jettisoned
0:09:14	Stage 2 engine cut-off (SECO)
0:09:49	Dragon separates from stage 2

Source: International Space Station partnership.

Table 4.1c On-orbit operations.

Time after lift-off, H:M:S	Event
0:11:53	Start sequence to deploy solar arrays
0:54:49	Demonstrate absolute GPS
2:26:48	Start guidance navigation control (GNC) bay door deployment (holds sensors necessary for rendezvous)
2:40:49	Relative navigation sensors checkout, checks the Light Detection And Ranging (LIDAR) and thermal imager
8:46:52	Demonstrate full abort, demonstrates Dragon's ability to abort with a continuous burn
9:57:58	Pulsed abort demonstration, checks Dragon's ability to perform abort using pulsating burns
10:37:58	Demonstrate Dragon's ability to free drift

Source: International Space Station partnership.

The next launch opportunity was three days away, assuming technicians could troubleshoot the problem. The technicians went to work and inspected the recalcitrant engine, determining that the issue had been a balky valve. After replacing the valve, a new launch attempt was announced for 3:44 EDT on 22 May.

C2+ LAUNCH

"The significance of this day cannot be overstated. It is a great day for America. It is actually a great day for the world because there are people who thought that we [NASA] had gone away, and today says, 'No, we're not going away at all.'"

NASA Administrator Charles Bolden

"Congratulations to the teams at SpaceX and NASA for this morning's successful launch of the Falcon 9 rocket from Cape Canaveral Air Force Station in Florida. Every launch into space is a thrilling event, but this one is especially exciting because it represents the potential of a new era in American spaceflight. Partnering with U.S. companies such as SpaceX to provide cargo and eventually crew service to the International Space Station is a cornerstone of the president's plan for maintaining America's leadership in space. This expanded role for the private sector will free up more of NASA's resources to do what NASA does best – tackle the most demanding technological challenges in space, including those of human space flight beyond low Earth orbit. I could not be more proud of our NASA and SpaceX scientists and engineers, and I look forward to following this and many more missions like it."

John P. Holdren, assistant to the President for Science and Technology

Preparations for the 22 May launch began late in the evening of 21 May as the tanking process started with the loading of 173,600 liters of liquid oxygen and 111,460 liters of kerosene. After fuelling was completed at T–2 hours 45 minutes, the countdown entered a quiet mode until T–1 hour, when Mission Control received its weather briefing, which showed an 80% chance of favorable launch conditions. At T–13 minutes, a final poll was conducted and, after a thumbs-up from all stations, the Terminal Countdown Sequence started at T–10 minutes. From that point, launch events proceeded normally. Three seconds before lift-off, the Merlin engines reached full thrust and moments later the hold-down system released the rocket. C2+ was finally on its way (Figure 4.8). The nine Merlins performed perfectly as the Falcon 9 arced upwards, passing maximum dynamic pressure 1 minute and 24 seconds following launch. Three minutes following lift-off, engine shutdown was completed and the first stage separated. The Merlin Vacuum engine took over by performing a six-minute burn and, after shutting down, Dragon was released into low Earth orbit (LEO).

SpaceX confirmed power generation was nominal and mission managers knuckled down in preparation for the shopping list of demonstrations and milestones that had to be completed before the vehicle could perform its rendezvous with the ISS. The first milestone had been completed with the deployment of the solar arrays (the first Dragon had relied on battery power). With its solar arrays operating normally, Dragon was in a LEO

4.8 Launch of C2/3. Credit: NASA

of 297 by 346 kilometers but it needed to change this in preparation for closing in on the ISS. The change in orbit was achieved by performing a co-elliptic burn that adjusted Dragon's orbit to 341 by 350 kilometers (inclination 51.66°). Once that had been achieved, Dragon opened its navigation pod bay door, exposing the vehicle's navigation instruments, which included a Light Detection And Ranging (LIDAR) range finder and thermal imager that were needed to lock on the ISS to provide range and velocity information. In addition to testing the LIDAR, Dragon's Star Trackers were also checked out, as was the absolute GPS system (AGPS) which was used to plot the vehicle's position in LEO.

One of the first checks was a pulsed abort test. Once Dragon had performed this successfully, the vehicle conducted a free drift demonstration to confirm to mission managers that the spacecraft could in fact enter free drift mode (FDM). This was one of the most important demonstrations because free drift was required during the final phase of station-keeping when the robotic arm would be used to grapple Dragon. With a successful free drift demonstration out of the way, Mission Controllers sat down to review the data in preparation for fly-under and fly-around demonstrations. The first place Dragon had to be was a point 2.5 kilometers below the ISS where Dragon intersected the outpost's R-bar. Getting to that point in space was achieved by performing height-adjust (HA) and co-elliptic (CE) burns. The first of these – the HA-2 burn – was fired early on 24 May. This was followed by the CE-2 burn less than an hour later. The two burns meant Dragon was now closing in on the ISS and communication was key. On board the orbiting outpost, Andre Kuipers and Don Pettit went to work powering up the UHF CUCU so that commands could be sent to Dragon during proximity operations. As Dragon became visible to the crew on board the ISS, a test strobe command was sent to Dragon using the ISS Crew Command Panel. This command activated a light on Dragon that confirmed to the ISS crew that the vehicle was in fact able to communicate. Also checked out during Dragon's approach was the DragonEye which acquired a thermal image of the ISS. Dragon crossed the station's R-bar 2.7 kilometers from the outpost before drifting away again as planned.

During its maneuvering, Dragon was restricted to the following:

- Closing (axial rate): 0.05–0.10 meters per second
- Lateral (radial) rate: 0.04 meters per second
- Pitch/yaw rate: 0.15° per second (vector sum of pitch/yaw rate)
- Roll rate: 0.40° per second
- Lateral misalignment: 0.11 meters
- Pitch/yaw misalignment: 5° (vector sum of pitch/yaw rate).

Now that the abort burns, AGPS checkout, and free drift recovery had been completed, it was time to start thinking about rendezvous. Key to this maneuver was the CUCU. The CUCU was designed to provide bi-directional communications between Dragon and the orbiting outpost using UHF ISS radio. The system was mounted on a rack on the ISS and was checked out on 25 January 2010 by Jeff Williams, Commander of Expedition 22. A few weeks later, on 11 March, Space X and NASA Mission Control conducted another series of tests that confirmed the communication link between ISS and NASA's Dryden station.

Rendezvous

Rendezvous operations began with a thorough check of Dragon's navigation systems. Once these systems were given the green light, Mission Control confirmed Dragon's position and velocity were accurate, and the approach began. Dragon made short engine pulses to start the rendezvous approach until it reached 220 meters on the R-bar. At this point, the ISS crew sent a retreat command to Dragon to make sure the capsule could retreat if any maneuver went pear-shaped. Once Dragon had completed the retreat maneuver, it fired its engines again to return to the 250-meter hold. Another check was performed at this stage by Mission Controllers who wanted to make sure Dragon's acceleration and braking were stable. Once Mission Controllers were satisfied, Dragon began its approach again before performing a second abort scenario. Once again, everything went well, and Dragon was cleared for close approach. Dragon fired its engines, and began gradually closing on ISS (see Tables 4.2 and 4.3). On board the ISS, Kuipers and Pettit kept a close eye on proceedings as Dragon entered the "Keep Out Sphere" (KOS). At 30-meters from the ISS, Dragon was held as Mission Controllers once again checked the capsule's status before conducting another Go/No Go poll. All systems were functioning flawlessly, and Dragon was given the go-ahead to creep towards the ISS until it was 10 meters from the station. Dragon had arrived at the capture point and free drift was initiated by disabling Dragon's thrusters.

With Dragon within reach of the station's robotic arm, all that remained was for Don Pettit to manipulate the Canadarm, grapple the capsule (Figure 4.9), and place the vehicle next to its berthing position. Once the Ready to Latch Indicators confirmed the spacecraft was in position for berthing, Dragon was hard-mated to the ISS and docked operations began.

Table 4.2 Fly-under timeline.

MET	Event
01/23:18	Height-adjust (HA) burn #2
02/00:04	Co-elliptic (CE) burn #2
02/00:15	Relative GPS demonstration
02/00:54	Entered ISS communication zone
02/02:44	Crossed R-bar at 2.5 km
02/03:23	Departure burn #1
02/04:10	Departure burn #2
02/06:47	Forward HA burn #1
02/07:33	Forward HA burn #2
02/12:14	Forward CE burn #2
03/16:28	Rear HA burn #1

Source: International Space Station partnership.

Table 4.3 Rendezvous timeline.

MET	Event
02/18:51	Rear height-adjust (HA) burn #2
02/19:37	Read co-elliptic (CE) burn #2
02/21:02	HA burn #2
02/21:48	CE burn #2
02/22:38	Entered ISS communication zone
02/23:16	HA burn #3
02/23:32	Mid-course correction #1
02/23:50	Mid-course correction #2
03/00:02	CE burn #3
03/00:27	Approach initiation burn
03/00:44	Mid-course correction #3
03/00:59	Mid-course correction #4
03/01:22	R-bar acquisition – range: 350 m
03/01:22	180° yaw
03/01:37	Range: 250 m – station-keeping
03/01:52	Retreat and hold demonstration
03/02:17	Range: 220 m – hold
03/02:32	Entered "Keep Out Sphere"
03/03:23	Range: 30 m – hold
03/03:37	Final approach
03/03:57	Range: 10 m – capture point
03/04:07	GO for Dragon capture
03/04:15	Capture
03/07:36	Berthing

Once berthing had been completed, the vestibule between Dragon and the ISS's Harmony hatch was checked for pressure leaks to ensure the seal was secure. Once that task had been performed, Mission Control gave the thumbs-up for hatch opening. Crewmembers then began outfitting the vestibule by installing ducts and sampling the air,

4.9 Dragon grappled by the station's robotic arm. Credit: NASA

Table 4.4 Undocking and re-entry timeline.

EDT	Event
12:35 a.m.	Dragon vestibule de-mate
12:50 a.m.	IPCU deactivation
01:35 a.m.	Dragon vestibule depressurization
04:05 a.m.	Canadarm detaches Dragon from Harmony nadir port
06:10 a.m.	Canadarm releases Dragon
06:11 a.m.	Departure burn 1
06:13 a.m.	Departure burn 2
06:20 a.m.	Departure burn 3
07:06 a.m.	Apogee reduction burn
07:07 a.m.	Unlatch/close/latch GNC cover
10:51 a.m.	De-orbit ignition
11:09 a.m.	Trunk jettison
11:28 a.m.	GPS blackout
11:35 a.m.	Drogue chute deploy
11:36 a.m.	Main chute deploy
11:44 a.m.	Splashdown

which were part of standard ingress tasks. With ingress operations completed, the crew began cargo transfer operations that included offloading the items listed in the Cargo Manifest and loading the items for the return trip (Appendix I). Following the exchange of cargo, Dragon's visit was almost complete. The vehicle was closed out, its hatch closed, the leak checks were conducted once again, and the vestibule was depressurized in preparation for undocking (Table 4.4). Using the Canadarm, Dragon was grappled to its release position 10 meters from the ISS. Dragon was once again in FDM with its engines disabled. Following checks of the vehicle's navigation instruments, Mission Controllers gave approval for release and Dragon was ungrappled. The robotic arm retreated, and Dragon reactivated its thrusters and performed three burns to begin its return to Earth.

De-orbit

Nearly five hours after being released by the Canadarm, Dragon performed its de-orbit burn and began its re-entry trajectory. Twenty minutes following the de-orbit burn, the vehicle was at the entry interface, and the PICA-X heat shield began to do its work, protecting the vehicle from temperatures that reached 1,600°C. As Dragon made its fiery descent, the vehicle stabilized its position using its Draco thrusters to control lift. At an altitude of 13,700 meters, Dragon's dual drogue chutes popped open and, at 3,000 meters, the main chute opened. Dragon's progress was tracked by a NASA P3 aircraft, which had also been used on some Shuttle missions. Splashing down (Figure 4.10) at 20 kilometers per hour about 450 kilometers off the California coast, Dragon's landmark flight came to an end following a mission elapsed time of 9 days, 7 hours, and 58 minutes. After being powered down, Dragon was transported to Los Angeles in preparation for its transport to Texas. Dragon's historic flight drew plenty of praise:

"Today marks another critical step in the future of American spaceflight. Now that a U.S. company has proven its ability to resupply the space station, it opens a new frontier for commercial opportunities in space and new job creation opportunities

4.10 Dragon after splashdown. Credit: NASA

right here in the U.S. By handing off space station transportation to the private sector, NASA is freed up to carry out the really hard work of sending astronauts farther into the solar system than ever before. The Obama Administration has set us on an ambitious path forward and the NASA and SpaceX teams are proving they are up to the task."

NASA Administrator Charles Bolden

"Congratulations to the SpaceX and NASA teams. There is no limit to what can be accomplished with hard work and preparation. This activity will help the space station reach its full research potential and open up space-based research to a larger group of researchers. There is still critical work left in this test flight. Dragon-attached operations and cargo return are challenging and yet to be accomplished."

William Gerstenmaier, associate administrator for NASA's
Human Exploration and Operations Mission Directorate

"There was reason to doubt that we would succeed because there wasn't a precedent for what we achieved. I think those reasons no longer remain having done what we have done so I hope those doubts are put to rest."

Elon Musk

A GIANT LEAP FOR COMMERCIAL SPACEFLIGHT

Dragon's trip to the ISS was a historic flight. Not only did the demonstration mission go a long way to dispelling the doubts that some members of Congress had expressed, it also confirmed that cargo transport to the ISS could be outsourced to commercial companies. The US government now had someone else they could call to ferry cargo up to the ISS. That confidence was demonstrated in August 2012 when NASA awarded SpaceX a US$440 million contract to develop a man-rated version of the Dragon to begin flights sometime in 2017. For the naysayers, the decision was a slap in the face. Those who had derided SpaceX as a company with too short a history to compete with the more experienced aerospace contractors were finally faced with the black-and-white reality: the traditional approach of top-down, sole-source, cost-plus contracting just wasn't viable any longer.

Dragon's demo flight to the ISS may have taken longer than expected but the results were well worth NASA's time and money. Having invested US$396 million, with a good deal of advice thrown in, NASA had made it possible for SpaceX to produce a real game-changer for less than what the agency spent on one suborbital launch of the Ares 1-X booster in 2009. The Dragon investment was also less than NASA had spent on the development of its Orion in the first half of 2012 alone. It didn't matter how you looked at the NASA–SpaceX deal, Dragon probably represents one of the best investments NASA made. Ever. Thanks to SpaceX, not only is the US back in the orbital transport business, but Dragon's stunningly low cost represents a rare bargain for taxpayers. Not convinced? Take a look at Orion. This Apollo-derived vehicle, which is built by Lockheed Martin, first flew in December 2014. One day, it will fly as part of the Space Launch System (SLS), which is expected to cost US$38 billion.[1] Read that again – US$38 billion. That's about 80 times the cost of the development of four commercial crew vehicles. Yes, Orion is being developed to fly to the Moon and is a little more complicated, but is it 80 times more complicated?

---

[1]   The number was quoted in the *Orlando Sentinel* on 5 August 2011. The *Sentinel* reported that NASA estimated the cost of the SLS and Orion Multi-Purpose Crew Vehicle (MPCV) could be as much as US$38 billion through 2021. The estimate, which came from an internal NASA report obtained by the *Sentinel*, stated that the cost of developing the SLS and the MPCV through 2017, the date of the first unmanned flight, was US$17–22 billion. Getting the vehicles ready for the first manned mission in late 2021 would be an additional US$12–16 billion.

5

Dragon Delivers

CRS1

With Dragon's demo flight out of the way, SpaceX's attention lasered in on the company's first Commercial Resupply Services mission: CRS1. As with so many commercial flights, CRS1 would be yet another milestone event. With a planned launch of 8 October 2012, SpaceX's Falcon 9 was rolled out at the end of August and put through a wet dress rehearsal (WDR) on 31 August. After the strong-back was lowered at T–100 minutes, fuel and thrust vector control bleeding was performed. This was followed at T–13 minutes by a flight readiness poll and a final hold at T–11 minutes. At T–4 minutes 46 seconds, the Falcon 9 was transferred to internal power and pressurization of the propellant tanks followed. The test concluded at the T–5 seconds mark with no engines being fired. After detanking, the rocket was rolled back in the hangar.

© Springer International Publishing Switzerland 2016
E. Seedhouse, *SpaceX's Dragon: America's Next Generation Spacecraft*,
Springer Praxis Books, DOI 10.1007/978-3-319-21515-0_5

Table 5.1 CRS1.

COSPAR ID	2012-054A
Launch date	8 October 2012, 00:34:07 Coordinated Universal Time (UTC)
Rocket	Falcon 9 v1.0
Launch site	Cape Canaveral SLC-40
Landing date	28 October 2012, 19:22 UTC
Regime	Low Earth
Inclination	51.6°
Berthing at the International Space Station	
Berthing port	Harmony nadir
RMS capture	10 October 2012, 10:56 UTC
Berthing date	10 October 2012, 13:03 UTC
Unberthing date	28 October 2012, 11:19 UTC
RMS release	28 October 2012, 13:29 UTC
Time berthed	17 days, 22 hours, 16 minutes
Cargo	
Mass	400 kg
Pressurized	400 kg

A couple of weeks later, it was announced that there had been a delay in the launch of the Soyuz TMA-06M vehicle due to a technical issue. The Soyuz had been scheduled to ferry three crewmembers (Kevin Ford, Oleg Novitskiy, and Evgeny Tarelkin) to the International Space Station (ISS) on 15 October, but this delay meant Dragon's CRS1 launch (Table 5.1) was much wider open, and it wasn't long before an amended launch date of 7 October (8:34 p.m. Eastern Daylight Time (EDT)) was announced.

Meanwhile, on orbit, gremlins were at work on the orbiting outpost's Robotic Work Station (RWS). Crewmembers use the RWS to control the Space Station Remote Manipulator System (SSRMS) from the Cupola as described in the previous chapter. In addition to the Cupola RWS, there is also a LAB RWS, and both systems needed to be operational during the docking. Unfortunately, the LAB RWS had suffered a power problem. Fortunately, thanks to the redundancy of the system, the LAB RWS was deemed functional and preparations for launch went ahead. Fortunately, CRS1 wasn't subject to the myriad delays of the Dragon demo flight, with launch (Figure 5.1) proceeding by the book at 20:35 EDT on 7 October. It was the fourth flight of the Falcon 9 and the ninth launch of a Falcon rocket (five Falcon 1 rockets were launched between 2006 and 2009, with two successes and three failures).

While the launch had proceeded according to plan, the flight wasn't without a few glitches when Engine #1 suffered an anomalous event about 80 seconds into launch. Fortunately, with nine engines, the Falcon 9 was able to deal with an engine failure and still complete the mission. Once in orbit, Dragon conducted its required series of burns to catch up with the ISS, where it was expected to arrive on 10 October. The arm grab was conducted by Akihiko Hoshide and Sunita Williams, and berthing operations proceeded smoothly with the opening of Dragon's hatch which was followed by ingress operations. Williams and Hoshide began transferring cargo (Appendix III) which included NanoRacks and some samples for the Microgravity Experiment Research Locker Incubator (MERLIN).

5.1   CRS1 launch. Credit: NASA

Once the cargo had been transferred, it was just a case of Dragon enjoying its time berthed to the ISS until its departure date of 28 October. Except for a network switch that required a reboot, Dragon's stay on orbit was uneventful. At the end of its three-week stay, the vehicle was loaded with life sciences samples and refrigerated bags and it de-orbited. As with the demo mission, Dragon splashed down a few hundred kilometers west of Baja, California, before being retrieved by barge for its trip to Long Beach.

> "Just a little over one year after we retired the space shuttle, we have completed the first cargo resupply mission to the International Space Station. Not with a government-owned and operated system, but rather with one built by a private firm – an American company that is creating jobs and helping keep the U.S. the world leader in space as we transition to the next exciting chapter in exploration. Congratulations to SpaceX and the NASA team that supported them and made this historic mission possible."
>
> *NASA Administrator Charles Bolden*

While Dragon's CRS1 mission had seemed to have gone without a hitch, there had been some anomalies that were identified post flight, one of which was the failure of one of the three computers while Dragon was berthed at the ISS. A radiation hit was suspected and the computer was restarted, although it could no longer synchronize with the other two computers. Radiation was also the suspected culprit in the failure of one of the three GPS units and the propulsion and trunk computers. After splashdown, three coolant pumps failed, which meant that the freezer was at −65°C and not the planned −95°C.

CRS2

Table 5.2 CRS2.

COSPAR ID	2013-010A
SATCAT no.	39115
Mission duration	25 days
Launch date	1 March 2013, 15:10 Coordinated Universal Time (UTC)
Rocket	Falcon 9 v1.0
Launch site	Cape Canaveral SLC-40
Landing date	26 March 2013, 16:34 UTC
Orbital parameters	
Reference system	Geocentric
Regime	Low Earth
Perigee	212 km
Apogee	326 km
Inclination	51.66°
Period	89.76 min
Epoch	1 March 2013, 16:55:51 UTC
Berthing at the International Space Station	
Berthing port	Harmony nadir
RMS capture	3 March 2013, 10:31 UTC
Berthing date	3 March 2013, 13:56 UTC
Unberthing date	26 March 2013, 08:10 UTC
RMS release	26 March 2013, 10:56 UTC
Time berthed	22 days, 18 hours, 14 minutes
Cargo	
Mass	677 kg
Pressurized	677 kg

SpaceX's second commercial mission (Table 5.2) to the ISS was set for 1 March 2013. Designated CRS2 and SpX-2, preparations for the mission began in November 2012 when the Falcon 9 was delivered to the SpaceX hangar for integration and checkouts. The Falcon 9 was followed by Dragon in December. On arrival at the Cape, the vehicle was integrated for launch – a process that included fitting the trunk section and testing the solar arrays. Cargo (Appendix IV) loading began in early February, shortly before an engine hot-fire test that confirmed that systems were functioning in advance of the full countdown rehearsal and Flight Readiness Review (FRR). One item that was subject to close scrutiny was the failure of Engine #1 during the CRS1 flight. In this event, Engine #1 had suffered a loss of pressure causing the Falcon's computer to initiate an engine shutdown. This in turn resulted in the rupture of the engine fairing. SpaceX claimed to have resolved the issue, although the details were not made public.

As launch date approached, SpaceX teams busied themselves inside the processing hangar, installing the nose-cone that protects the vehicle's Common Berthing Mechanism (CBM) during launch. Once final checks had been completed, the integrated launch vehicle was rolled to the pad, where the transporter placed it in the vertical position. Electrical cables and propellant umbilicals were then attached in readiness for the countdown rehearsal. Since SpaceX was so confident with their work, the WDR and static fire, which were normally separate events, were combined. All countdown events proceeded by the book and concluded with the nine Merlin 1C engines reaching lift-off thrust for two seconds before being shut down.

After reviewing the WDR, it was time for the launch readiness test (LRT), which approved the launch. At the time of the LRT, there was some concern about gusty northwest winds, but final launch preparations continued with the goal of launching on time. Fifteen hours before launch, Dragon began a late cargo load – a process that included experiment materials that were to ride inside the Glacier Laboratory Freezer and fresh food. Seven and a half hours before launch, the Falcon 9 was powered up and technicians prepared for propellant loading. At T–13 minutes, a Go/No Go decision was polled for terminal count and, at T–10 minutes, the Falcon 9 was in its final pre-launch configuration. At T–7 minutes, Dragon began the transition to internal power and the Falcon 9's computers aligned for launch. Everything was proceeding by the book. At T–3:11, the Flight Termination System was armed and final status checks were performed by the range. Shortly after the thrust actuator test at T–50 seconds, the propellant tank pressurization began and, three seconds before launch, the Falcon 9's Merlin engines roared into life. On reaching lift-off thrust of 430,000 kilograms, CRS2 blasted off. Dragon was once again on its way to the ISS. Everything seemed to have gone like clockwork but, on reaching orbit, SpaceX declared an anomaly.

Dragon had been placed in a 199 by 323 kilometer orbit with an inclination of 51.66° (the planned orbit had been 200 by 325 kilometers). The first hint that something was awry was two minutes after spacecraft separation when Dragon was supposed to have begun its deployment of the solar arrays. The problem stemmed from a balky propellant valve (see sidebar) that had decided to go into passive abort mode, which meant SpaceX engineers had to try to reconfigure their computers to override the system. After some finagling, the solar arrays were deployed but the time taken to troubleshoot the issue had meant that Dragon had been unable to make its co-elliptic (CE) burn (these were required to

Table 5.3 Dragon SpX-2 rendezvous timeline.

Coordinated Universal Time (UTC)	Event
03:23	Height-adjustment (HA) burn
04:08	Co-elliptic (CE) burn
06:34	ISS maneuver to comm. attitude
	Duration: ~10 min
06:40	Dragon enters ISS comm. zone
07:15	HA burn
07:31	Mid-course correction 1
07:48	Mid-course correction 2
08:01	CE burn
08:30	Approach initiation burn
08:46	Mid-course correction 3
09:03	Mid-course correction 4
09:07	ISS maneuver to capture attitude
09:23	R-bar acquisition – range: 350 m
09:23	Hold – 180° yaw maneuver
09:28	Resume approach
09:37	250-m hold
09:47	Resume approach
10:39	30-m hold
10:53	Resume approach
11:11	Capture point arrival – range: 10 m
11:21	MCC – GO for capture
11:31	Capture
FD3	Berthing operations

circularize the vehicle's orbit) as scheduled: this was because the vehicle was programmed not to deploy the solar arrays until Dragon had reached its correct attitude. This in turn meant that the vehicle would be unable to rendezvous with the ISS that day.

Fortunately it wasn't long before Dragon was pronounced operational again and NASA and SpaceX put their heads together to figure out a revised rendezvous profile (Table 5.3). After some discussion, a new rendezvous date of 3 March was announced and Dragon began its orbit adjustment and phasing maneuvers that boosted its orbit to a 315 by 341 kilometer orbit. Two more height-adjustment (HA)/CE burns followed that placed the vehicle into a 394 by 406 kilometer orbit. Another burn was completed that circularized the orbit at 440 by 453 kilometers.

At this point, Dragon was flying 2,000 kilometers ahead and above the orbit of the ISS, so it was simply a case of waiting for the station to pass underneath, at which point Dragon fired its engines to cross the V-bar to set up the rendezvous. After being green-flagged for rendezvous, Dragon performed the requisite orbital maneuvers, which included a HA/CE burn that placed the vehicle inside the station's communication zone. Relative GPS and UHF communications were established and Dragon made its way to the point 2.5 kilometers below and behind the station (see Chapter 4 for details). Another HA maneuver was performed that increased Dragon's altitude and a couple of mid-course correction burns were performed to fine-tune Dragon's trajectory before a CE burn was executed to place the pride of SpaceX 1.4 kilometers below and behind the station. Then it was just a case of Dragon performing its approach initiation burn to place the vehicle en route to intercept the outpost's R-bar. Business as usual. On board the ISS, Kevin Ford, Tom Marshburn, and Chris Hadfield kept a close eye on Dragon as it approached to within 1,000 meters from the station. As with CRS1, the crew observed from the Cupola and used the RWS and Crew Command Panel (CCP) to make sure the rendezvous progressed nominally. Once Dragon had acquired the R-bar, it executed a 180° yaw maneuver orienting the vehicle so that its nose was facing away from the station in case an abort was required. At the 250-meter hold point, Dragon's navigation instruments were verified to ensure they were providing accurate data. Once the instruments had been checked out, Dragon began its final approach as it entered the 200-meter "Keep Out Sphere" (KOS). Dragon's travel time from 250 meters to 30 meters from the station took 44 minutes. The reason for such slow progress was because the KOS zone is one where there are extremely high safety requirements and Mission Controllers need to perform continuous status checks to ensure all systems are functioning as advertised. Dragon arrived at capture point at 10:22 UTC, at which point Mission Controllers conducted a final check of Dragon's navigation sensors before the vehicle's free drift mode. Kevin Ford, who was operating the Canadarm, maneuvered the arm to position the Latching End Effector (LEE) above the grapple fixture in preparation for the capture sequence. Dragon was captured nine minutes later. It had been a flawless rendezvous:

"Let me just say congratulations to the SpaceX and Dragon team in Houston and in California. As they say, it's not where you start but where you finish that counts. You guys really finished this one on the mark. You're aboard, and we've got lots of science on there to bring aboard and get done, so congratulations to all of you."

Expedition 34 Commander Kevin Ford

Once Dragon was berthed at the Harmony module (Figure 5.2), the crew went to work taking images using the Node 2 Centerline Berthing Camera to send to the ground to ensure the pressure seal between the vehicle and the station was secure. As Chris Hadfield operated the CBM, the rest of the crew were positioned inside Harmony ready to secure Dragon to the ISS. This process required two stages: the first-stage capture was complete after four latches were closed, at which point the robotic arm was powered down, setting the stage for second-stage capture which was completed once four sets of four bolts secured the vehicle in position. Once all these procedures had been completed, the clock showed 13:56 UTC. Dragon was now officially part of the ISS, where it was due to remain for three weeks. After leak checks had been completed, Mission Controllers gave the go-ahead to open the Harmony hatch and the crew proceeded with fitting out the vestibule. This was quite an involved procedure that required the removal of Harmony's thermal cover and the de-mating of the control panel assemblies which had been used to fix the bolts. Once the vestibule had been fitted, the crew opened Dragon's hatch at 18:14 UTC and performed air sampling tests and checked Dragon's interior for debris before leaving the vehicle for 20 minutes so that the air inside the vehicle could mix with that inside the ISS. With a cargo of 1,050 kilograms, there was plenty of unloading work to be done.

5.2   Dragon at the International Space Station (ISS). Credit: NASA

5.3 Dragon unberthing. Credit: NASA

And once all the cargo had been unloaded, there was the task of stowing 1,370 kilograms of cargo for the return trip. On its previous visits to the ISS, Dragon had only delivered pressurized cargo, but this time the vehicle had carried unpressurized cargo in the trunk section (the first two Dragons had flown with empty trunks).

After three weeks as a guest attached to the ISS, Dragon was loaded with return cargo and prepped for return to Earth. Due to inclement weather at the splashdown site, the vehicle remained on orbit for an extra day, which allowed Tom Marshburn and Chris Hadfield some more time to practice the unberth and release procedures. Prior to departure, Hadfield and Marshburn had to configure the vestibule that required installing four control panel assemblies to drive the bolts that had secured Dragon in place. Once the bolts were unbolted, the choreography of de-mating began after a visit of 22 days, 18 hours, and 14 minutes. With Dragon in its release position (Figure 5.3), the crew checked the vehicle's navigation instruments and Mission Controllers fine-tuned the vehicle's sensors and thermal imagers prior to the vehicle being placed in free drift. With its thrusters inhibited, the ungrappling sequence began with Dragon being released at 10:56 UTC. One minute following release, Dragon made its first burn followed by a second three minutes after release that powered the vehicle outside the KOS. At this point, NASA's responsibility ended and authority for the mission was passed over to SpaceX (Table 5.4).

After a few hours in free flight, Dragon executed a trim maneuver that trimmed its orbit and lowered its apogee. This event was followed by closing the GNC bay doors. At 15:42 UTC, Dragon executed its de-orbit burn that placed the vehicle on its re-entry trajectory. Nineteen minutes after the de-orbit burn had been performed, Dragon's trunk was

Table 5.4 Return timeline (26 March 2013).

GMT	Event
08:05	Unberthing
10:56	Dragon release
10:57	Departure burn 1
10:59	Departure burn 2
11:03	180° yaw
11:06	Separation burn
11:07	Depart "Keep Out Sphere"
~11:52	Orbit-adjust burn
~11:53	Guidance navigation control (GNC) bay door closure
15:42	De-orbit burn
	Duration: 9–10 min – Delta-V: ~100 m/s
16:01	Trunk jettison
~16:16	Entry interface
16:20	GPS blackout
16:27	Drogue chute deploy
16:28	Main chute deploy
16:34	Splashdown

jettisoned and the vehicle used its thrusters to stabilize its position through re-entry. Splashdown occurred at 16:34 UTC, 385 kilometers off the California coast. It had been another successful mission, but it's worth noting that Dragon wasn't the only commercial cargo vehicle. While Dragon was executing its flight, Orbital Sciences Corporation had been busy prepping their inaugural mission of their Antares and Cygnus spacecraft (Figure 5.4). At the time of the CRS2 mission, Orbital had an April launch date in mind for the Antares and July launch for the COTS demo flight of the Cygnus. If all proceeded as planned, the first Commercial Resupply Services (CRS) Cygnus flight would take place in October (it actually launched on 13 September 2013).

And Cygnus and Dragon weren't the only ISS cargo flights. The European Space Agency's (ESA) Automated Transfer Vehicle (ATV) was also scheduled to fly (Figure 5.5) that summer, as was Japan's H-II Transfer Vehicle (HTV-4) (Figure 5.6) and the Russian Progress vehicle (Figure 5.7) that was scheduled for a November flight. With five resupply spacecraft on tap, the world's space agencies were all of a sudden spoilt for choice when it came to delivering cargo to the ISS, but so many vehicles caused a bit of a headache for the ISS flight manifest and scheduling.

Dragon the Weapon

Dragon's thruster glitch after arriving on orbit highlighted the strange world of arms regulations because, under US law at the time, Dragon was classified as a weapon. Yes, you read that correctly. Under the International Traffic in Arms Regulations (ITAR), Dragons are listed as munitions. Why? Well, since 1999, ITAR has prevented North Korea, China, and Iran from getting their hands on dual-use technology, and Dragons fit that designation. Strange but true.

5.4 The Cygnus spacecraft. Cygnus is an unmanned vehicle operated by Orbital Sciences Corporation. Just like Dragon, Cygnus is in the business of ferrying cargo to the International Space Station (ISS). The program was started as part of NASA's Commercial Orbital Transportation Services Program. The vehicle is launched by the Antares rocket from the Mid-Atlantic Spaceport in Virginia. Unlike Dragon, Cygnus can't return cargo to Earth and simply burns up on re-entry. The standard Cygnus can fly up to 2,000 kg of cargo, although an enhanced version of the spacecraft will be able to transport up to 2,700 kg. Credit: NASA

CRS3

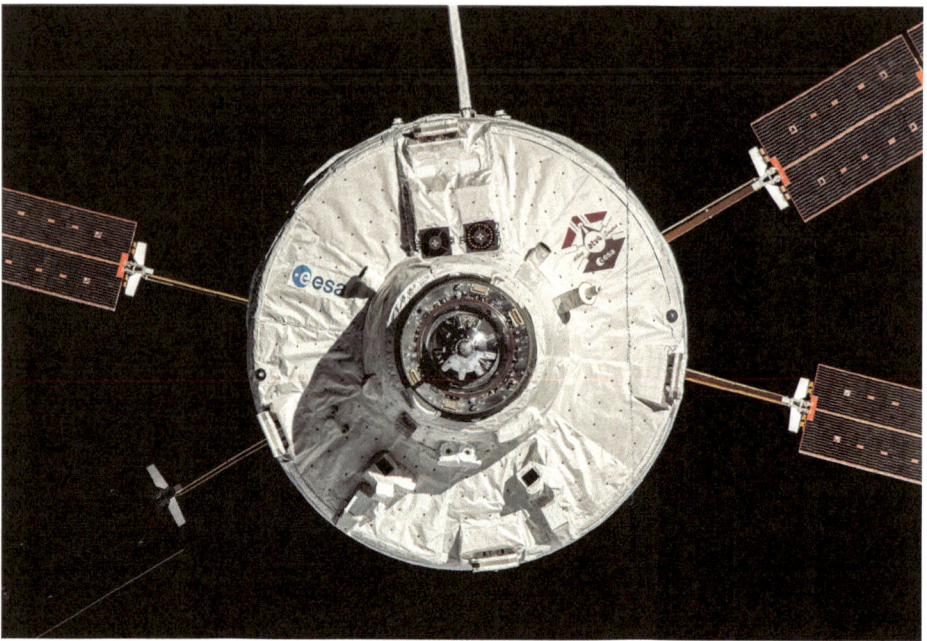

5.5 The Automated Transfer Vehicle (ATV) is a European Space Agency (ESA) vehicle that was designed to resupply the International Space Station (ISS). In addition to carrying up to 7,500 kg to the orbiting outpost, the ATV can also be used to reboost the ISS. The ATV is launched by an Ariane 5 ES launcher from French Guiana. Once in orbit, it docks with the Zvezda Service Module where it can stay for up to six months. In common with the Cygnus, the ATV cannot return cargo from the ISS and performs a destructive re-entry after undocking. Credit: NASA

One of SpaceX's primary goals is to make the Falcon 9 a reusable launch vehicle and one of the first steps towards realizing that goal was to test four landing legs during the CRS3 mission. The plan was to deploy the legs after the first-stage separation and execute a soft splashdown. At the time, testing had been conducted using the Grasshopper (Figure 5.8), with a long-term goal of incorporating the touchdown technology into Dragon so that, one day, astronauts returning from the ISS can make a powered landing. Launch was scheduled for 16 March 2014 at 08:41 UTC. Thanks to the bigger boost of the Falcon 9 v1.1, Dragon would be carrying 1,580 kilograms (Appendix V) into orbit. The static fire was conducted without issue but, as launch day approached, contamination (caused by machine oil used in an industrial sewing machine) was discovered in the beta cloth shields in the trunk, causing a delay of two weeks with a revised launch date of 30 March. Then, on 27 March, another delay was announced due to a range outage and damaged radar. Yet again, the launch date was revised to 21 April but, on 20 April, another glitch threatened a delay. This time, the problem was on board the ISS and the offending item was one of the

5.6 Japan's H-II Transfer Vehicle (HTV) is used to deliver cargo to the International Space Station (ISS). Built and operated by the Japanese Space Agency (JAXA), the HTV is launched by the H-IIB launcher. Once berthed with the station, it can remain there for up to a month. In common with the Cygnus and ATV, the HTV performs a destructive re-entry. Credit: NASA

myriad multiplexers/demultiplexers (MDM). While not a serious problem for the ISS crew, the MDM issue was inextricably linked to Dragon because it provided redundancy for control of the Canadarm. The ISS crew tried rebooting the MDM without success, which prompted mission managers to suggest a contingency spacewalk. Ultimately mission managers decided to delay the spacewalk until after the CRS3 launch, which was now set for 21 April at 20:58 UTC.

But it wasn't to be. A helium leak was detected on the Falcon 9's first stage (in the pneumatic stage separation system) and the launch was scrubbed. Incidentally, helium is used to pressurize the propellant tanks in flight, to spin the turbo pumps of the Merlin engines, and to deploy the landing legs. As engineers safed the vehicle, mission managers mulled over the next launch attempt and decided on 25 April at 19:25 UTC. Meanwhile, ISS mission managers were still trying to work the MDM spacewalk into the schedule. The spacewalk, which had originally been slated for 22 April, would be performed by Rick Mastracchio and Steve Swanson, who would replace the balky MDM with a new one. But, with the launch scrub, mission managers now had to rework the schedule to allow sufficient time after Dragon's arrival. While mission managers worked the schedule on the ground, Rick Mastracchio and Steve Swanson busied themselves for the spacewalk

5.7 The Progress has been flying for decades. Its first flight was way back in 1978 when one flew to the Russian Salyut 6 space station. Over the years, the vehicle has been upgraded several times. Its most recent designation is the Progress M-M. It is launched by a Soyuz rocket and it can dock with any module of the Russian segment of the International Space Station (ISS). The Progress launches without a crew but, once docked with the station, the crew enter the vehicle, which means the Progress is classified as a manned spacecraft. Three or four Progress vehicles visit the station every year. In common with the ATV, Cygnus, and HTV, the Progress cannot return cargo. Credit: NASA

5.8 Grasshopper in action. Credit: NASA

by preparing extravehicular activity (EVA) suits EMU-3005 and EMU-3011 (which, for those history buffs, was the suit worn by Luca Parmitano during the water intrusion event in July 2013: ISS EVA-22).

Back on the ground, SpaceX readied for the 25 April launch. This time, nothing prevented the Falcon 9 v1.1 launching on schedule, and the rocket gave Dragon a smooth ride to orbit. On board was an assortment of supplies, ranging from spare legs for Robonaut, food, a spacesuit, and myriad scientific payloads. After completing its far-field phasing maneuvers and vehicle checkouts, Dragon prepared for its rendezvous and robotic capture, which was planned for 27 April at 11:14 UTC. Capture occurred as planned and Dragon was docked to the Harmony module at 14:06 UTC. Except for a minor issue with an isolation valve in its propulsion system during its orbital maneuvers on 25 April, the flight had gone very smoothly. The crew contacted Mission Control saying they wanted to open the hatch that same day but one of the CBM bolts (Bolt 3–1) proved troublesome. Fortunately, after some troubleshooting, the offending bolt retracted: incidentally, only 14 of the 16 bolts are needed to hold pressure, so 15 would have done the job. With the bolt issue resolved, the crew moved on to leak check operations but, by the time the ventilation ducts had been readied, it was getting late and the crew decided against working overtime, deciding instead to open the hatch the next day.

After the hatch was opened on 1 May, one of the first tasks was to use the robotic arm to move two external payloads from Dragon's trunk to their operational locations. One of the payloads was the High Definition Earth Viewing (HDEV) payload (Figure 5.9) that

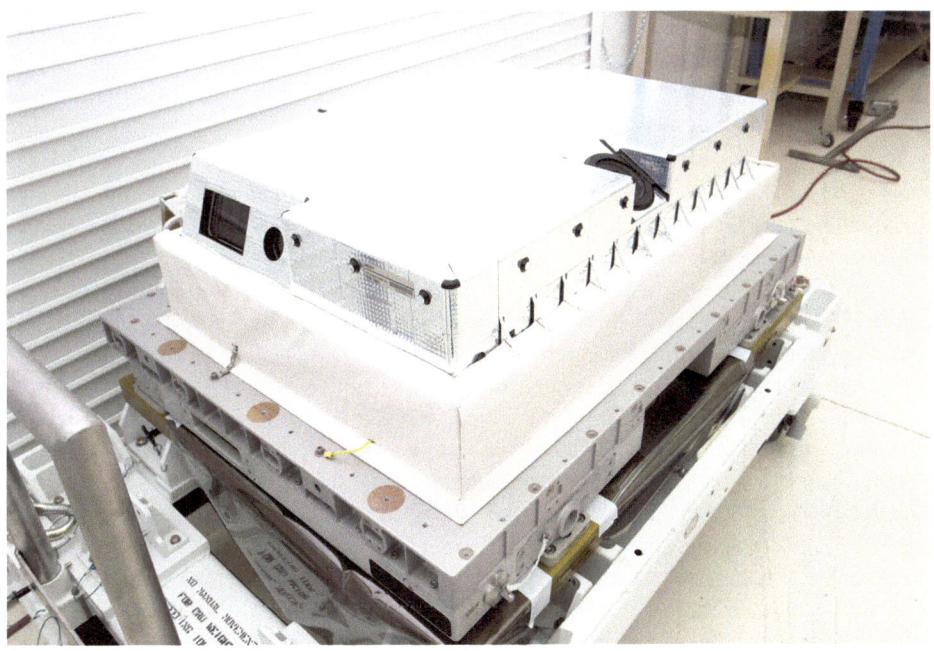

5.9 High Definition Earth Viewing (HDEV) equipment. Credit: NASA

5.10   NASA/JPL's Optical Payload for Lasercomm Science: OPALS. The laser beam you can see in the image will be transmitted from a ground telescope to OPALS as part of the demonstration of optical communication in space missions. Laser beams can support high data rates of up to 50 megabits per second. This means optical communication could benefit deep space missions which usually rely on 200–400 kilobits per second. Credit: NASA

was to be fitted to the External Payload Facility of the ESA's Columbus module. The other payload was the Optical Payload for Lasercom Science (OPALS), which was designed to demonstrate laser communication from orbit (Figure 5.10).

In preparation for moving HDEV and OPALS, the Mobile Transporter was repositioned from Worksite 2 (WS2) to WS6 and the robotic arm was moved to a position where it could grapple the Special Purpose Dexterous Manipulator (SPDM), aka Dextre (Figure 5.11). Once Dextre was ready, Canadian controllers worked with Mission Controllers in Houston to move the robotic handyman to a location where it could inspect Dragon's trunk. All systems were ready on 3 May when Dextre was powered up in preparation to remove the HDEV from Dragon's trunk. By using cameras inside the trunk and cameras on Dextre, the robot was guided inside the trunk where it placed its ORU/Tool Changeout mechanism on the HDEV grappling interface. With a solid grip on its payload, the HDEV's Flight Releasable Mechanism was released in preparation for the payload's transfer to the Columbus module. Here the HDEV was positioned on the Earth-facing (nadir) location. Once in position, the necessary electrical and data interfaces were hooked up and the

5.11   Dextre manipulating the HDEV equipment. Credit: NASA

HDEV was integrated into the ISS systems. Now it was time to transfer the OPALS, which was to be installed on the Express Logistics Carrier 1. This required a slightly more complex sequence of movements.

The day after HDEV was installed, Canadian Space Agency (CSA) Mission Control in Saint Hubert, Quebec, got busy moving the Canadarm 2 and Dextre to a position where it could grab OPALS. Once again, using the ORU/Tool Changeout Mechanism, Dextre tried to grapple OPALS, but a minute offset (0.005 centimeters) meant the robot couldn't get a good grip. After new payload files were written and uploaded to the ISS, the task resumed. With the recalibrated software, Dextre captured OPALS which was then positioned on the Enhanced ORU Temporary Platform (EOTP). From the EOTP, Canadarm-2 moved Dextre to a position where it could begin the sequence of movements that ended with OPALS being installed on the P3 truss of the ISS. As with HDEV, OPALS was connected with electrical and data cables, which marked the end of external cargo operations for CRS3.

While HDEV and OPALS were being retrieved and positioned, the crew unpacked the rest of Dragon's cargo and began loading almost 1,600 kilograms of cargo for return to Earth. On 18 May, Steve Swanson used the robotic arm to release Dragon at 13:26 UTC.

At 18:12 UTC, Dragon executed its de-orbit burn, splashing down in the Pacific Ocean 490 kilometers west of Baja, California, 53 minutes later to close out a very successful mission. After time-critical cargo had been shipped to NASA, SpaceX began investigating a water intrusion that occurred while Dragon was floating in the ocean. Although no loss of science had occurred, it was the second time a water intrusion event had occurred: on CRS1, water had entered the avionics compartment, which prompted SpaceX to implement a temporary fix for CRS2. Initially it was thought a water sample bag had ruptured, but none of the bags had failed so it was possible that water had entered through a pressure equalization valve.

CRS4/SpX4

Ambitious as ever, SpaceX planned its fastest turnaround between missions as it planned a 23 September 2014 launch date for its CRS4 flight. It was ambitious because SpaceX's Cape Canaveral facility was limited when it came to parallel processing and the CRS4 flight was planned just 13 days after a Falcon 9 flight was due to launch the AsiaSat-6 satellite. Another limiting factor was SpaceX's ground support systems, but SpaceX reckoned they were able to adjust this to fit the 23 September launch date. The Dragon SpX-4 vehicle arrived at Cape Canaveral on 9 July and began its processing operation (Table 5.5). Two weeks later, Dragon's trunk arrived and preparations began to install two payloads: a nadir-viewing adapter for the Columbus External Payload Facility and the RapidScat payload which will be discussed later. Ferrying an upmass of 2,272 kilograms and a downmass of 1,734 kilograms (Appendix V), SpX-4 would set yet another cargo record. On 7 September, SpaceX launched the AsiaSat-6 from Launch Complex 40 – a launch that marked the 12th consecutive successful launch of the Falcon 9. Then, on 18 September, SpaceX conducted a static fire test of the SpX-4 Falcon 9, setting the stage for a 20 September launch date, pending favorable weather. The problem was an upper-level

Table 5.5 SpX-4.

Time	Event
T–00:03	Merlin 1D engine ignition
T–0	LIFT-OFF
T+00:1x	Pitch and roll
T~01:10	Mach 1
T~01:17	Maximum dynamic pressure
T~01:45	Merlin 1D vac engine chilldown
T+02:41	Main engine cut-off
T+02:44	Stage separation
T+02:52	Stage 2 ignition
T+03:32	Dragon nose-cone jettison
T+08:55	Terminal guidance mode
T+09:05	Flight termination system safing
T+09:40	Stage 2 cut-off
T+10:15	Dragon separation
T~11:00	Solar array deployment
Target orbit: 200 by 360 km, 51.6°	

trough that threatened rain showers and thick clouds. The problem with the CRS4 launch was that the Falcon 9 had an instantaneous launch window, which meant there was no margin for delay.

Unfortunately, that was exactly what happened, and SpaceX had to scrub the launch after thunderstorms moved into the launch area. Mission Controllers studied the next day's weather which had a 40% chance of favorable conditions and planned for a launch at 05:52:04 UTC. Fortunately, the weather on 21 September cooperated and the Falcon 9 carried Dragon and its assorted cargo into orbit for a planned four-week stay at the ISS. In what had now become almost routine, Dragon spent the first day on orbit undergoing extensive checks, while on board the ISS crewmembers prepped the Canadarm 2.

On 22 September, Dragon was en route to the ISS, where Alexander Gerst and Reid Wiseman were preparing for docking operations. The next day, Dragon made a punctual arrival at the ISS after being captured by the robotic arm at 10:52 UTC. After performing a flawless rendezvous, Dragon was once again parked at the Harmony module and was soon primed for business after the capture latches were secured at 13:21 UTC. After taking air samples and installing the Intermodule Ventilation Equipment, Gerst, Wiseman, and Maksim Suraev exited Dragon to allow air to circulate, before returning to begin the process of unloading 1,627 kilograms of pressurized cargo (589 kilograms was being carried in the unpressurized trunk). One interesting item was the RapidScat (see sidebar) instrument that was due to be attached by the Canadarm-2/Dextre to the External Payload Facility of the Columbus Module. Other payloads of interest included the Rodent Habitat and the Rodent Transporter and Access Unit (see sidebar) and a 3D printer – the first in space.

RapidScat

RapidScat is a scatterometer (Figure 5.12) that will be used to measure surface winds over the ocean. Specifically, RapidScat will provide data on diurnal wind variations, which is important information because these variations are not well understood and they have a significant impact on how clouds form in the tropics.

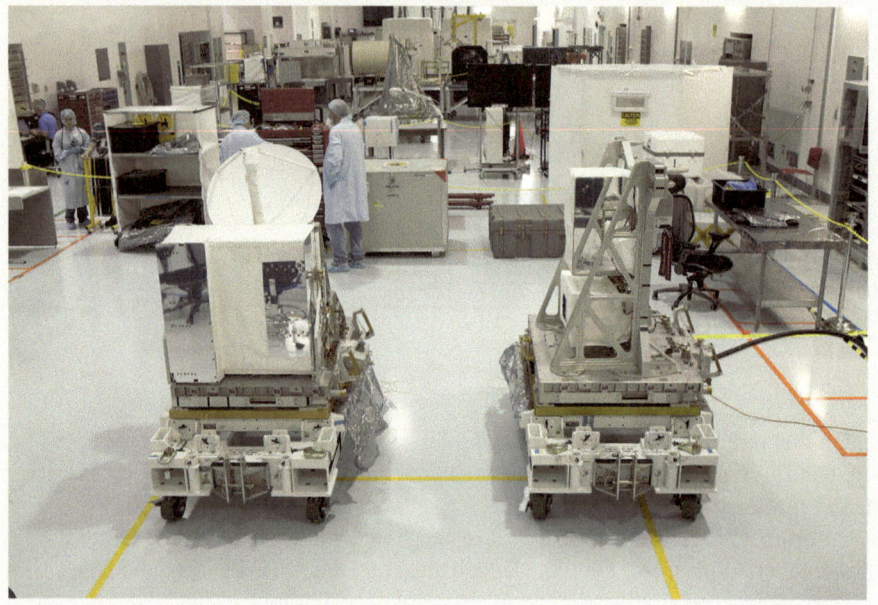

5.12 The RapidScat payload waits to be configured for launch. The RapidScat will be used to monitor ocean winds and hurricanes. Credit: NASA

Rodent Habitat

The Rodent Habitat (Figure 5.13) needs very little explanation, but is mentioned here to highlight the fact that there are plenty of experiments still being performed on orbit that require the services of mice and rats. The habitat can support as many as 10 mice or six rats. A key support element to the habitat is the Rodent Transporter which is used during the transport of the research subjects to orbit. The transporter features everything a mouse or rat could want – except an escape option. There is plenty of water and food bars are installed that support up to 20 mice for more than a day.

(continued)

(continued)

It also features its own life-support system which ensures fresh air is circulated inside the habitat and contamination is prevented from entering the habitat from the ISS by the use of filters. One option that isn't available to the test subjects is privacy because their movements are tracked 24/7 thanks to myriad cameras and infrared imagers. Another element is the Animal Access Unit (AAU), which interfaces with the transporter, habitat, and microgravity science glovebox. Here the crew can manipulate the mice and/or rats. Why would they want to do this? Well, when it comes to returning the test subjects to Earth, the first step is usually euthanization by means of formalin. These guys pay the ultimate price when it comes to science.

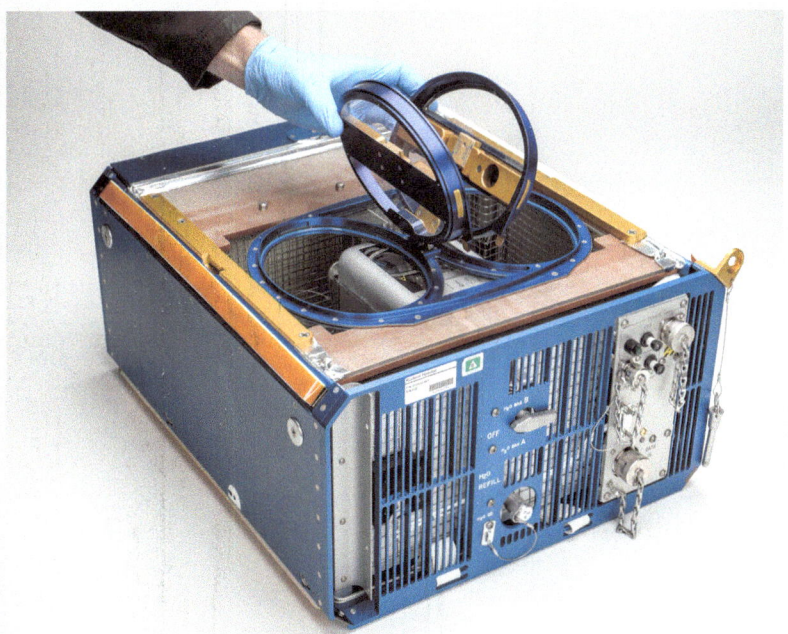

5.13 The Rodent Habitat. This is a multipurpose system that features life support, an access unit, and the transporter which doubles as housing for the rodents once they have arrived on the station. The habitat can support 10 mice for up to 90 days. Depending on the Principal Investigator's requests, the crew can perform dissections, blood collection, bone densitometry, and tissue preservation. Credit: NASA

3D Printer

Manufactured by Made in Space Inc., the 3D printer (Figure 5.14) delivered by SpX-4 is the first such printer in space where it will serve as a kind of machine shop on orbit. Incidentally, SpaceX has been using 3D printing for a while and used the technology to print its SuperDraco engine. The ISS printer uses acrylonitrile butadiene styrene (ABS) resin to create components that are also created on Earth using the identical printer. Once a number of components have been manufactured, scientists will examine the duplicated items to see whether there is any difference between those manufactured on orbit and those made on Earth.

5.14 The Made in Space 3D Printer being checked out in a parabolic flight. Credit: NASA

Dragon also delivered a satellite deployment mechanism with the acronym SSIKLOPS. Until the arrival of SSIKLOPS, the only satellites that could be deployed were CubeSats but, with SSIKLOPS on board, satellites of all shapes and sizes should be deployable. On 25 October, after a lengthy stay on orbit, Dragon closed out its stay after its return to Earth had been postponed by three days due to high seas in the recovery area. Dragon was released at 13:57 UTC and the vehicle made its way down the R-bar before making its final departure burn that sent the spacecraft on its way for good. After a few hours in free flight, Dragon executed its trim maneuver to lower its altitude and set itself up for the de-orbit burn which was performed at 18:43 UTC. Splashdown occurred 56 minutes later. In what was becoming almost routine, time-critical cargo was shipped to NASA while the rest of the cargo was unloaded once Dragon arrived back in Texas.

CRS5/SpX5

Emboldened by the run of Dragon's mission successes, SpaceX were on a roll in 2014 and the fifth commercial flight was scheduled for 17 December, less than two months after the return of SpX-4/CRS4. Much of the talk about SpX-5 (Table 5.6) was not about Dragon, but about the test of SpaceX's reusable rocket and the plan to return the first stage to a platform floating in the Atlantic Ocean. The hardware to achieve this ambitious goal had been tested using the Grasshopper in 2012 and then in September 2013 using the Falcon 9 v1.1 launcher: in this mission, the first stage performed a braking burn and executed a single-engine landing burn before splashing down in the ocean. The next test of the landing system was performed during the SpX-3 mission when the first stage soft-landed in the ocean. This was followed by another test during the SPx-4 mission when the first stage performed another soft landing/splashdown. The next step after these successful missions was to attempt a landing on solid ground, hence the use of the floating platform, or Autonomous Spaceport Drone Ship to use the correct designation. The plan for the mission was straightforward: the Falcon 9 would first deliver the launcher's second stage to the trajectory required for orbit after which the separated first stage would continue to an apogee of about 140 kilometers where it would execute a boost-back maneuver and prepare to target the landing platform. Then, as the first stage entered the atmosphere, it would perform a supersonic retro propulsion burn between 70 and 40 kilometers altitude, before moving towards the landing platform where it would slow its vertical velocity to two meters per second. Finally, it would deploy four landing legs and land on the platform. Piece of cake.

Lift-off was set for 18:22 UTC on 19 December, but this was delayed to no earlier than 6 January 2015 after the Falcon 9's static fire failed to meet all the test objectives. This meant that the ISS crew had to wait a little longer for the Christmas treats, most of which were due to be ferried up by Dragon. It was the second delay after the first launch had been postponed by a week to 16 December at NASA's request following the Antares failure on 28 October. The Antares launch failure was the third operational launch of the Cygnus cargo vehicle and, when that went up in flames, some of the cargo items were substituted

Table 5.6 SpX-5.

Time	Event
T–00:03	Merlin 1D engine ignition
T–0	LIFT-OFF
T~00:18	Pitch and roll
T+01:10	Mach 1
T~01:17	Maximum dynamic pressure
T~01:45	Merlin 1D vac engine chilldown
T+02:xx	Stage 1 throttle segment
T+02:37	Main engine cut-off
T+02:41	Stage separation
T+02:49	Stage 2 ignition
T+03:05	Stage 1 maneuver out of stage 2 plume
T+03:29	Dragon nose-cone jettison
T~04:00	Stage 1 apogee (140 km)
T+04:35	Stage 1 boost-back burn
T~05:00	Stage 1 grid fin deployment
T+07:19	Stage 1 landing burn
T+07:40	Stage 1 completes landing burn
T+08:45	Terminal guidance mode
T+08:55	Flight termination system safing
T+09:27	Stage 2 cut-off
T~09:35	Stage 1 landing leg deployment
T~09:45	Expected stage 1 landing
T+10:02	Dragon separation
T~11:00	Solar array deployment

on Dragon to make sure the ISS crew had all the supplies they needed. The reason the SpX-5 launch had to be pushed all the way back to no earlier than 6 January was due to the orbital plane (beta angle) of the ISS: between 28 December 2014 and 7 January 2015, the orbiting outpost would be in almost continuous sunlight, which would result in high thermal loads that would prevent vehicles from visiting.

So launch plans were recalibrated for a 6 January launch, but this was aborted due to a problem with the thrust vector control actuator on one of the engines of the Falcon 9's second stage. Pending an investigation, SpaceX announced a new launch date of 9 January, but this was pushed to 10 January when Mission Controllers realized there just wasn't enough time to complete all the technical reviews.

Finally, on 10 January 2015, the SpX-5 mission was underway as the Falcon 9 arced upward from Space Launch Complex 40 at 09:47 UTC. At about the same time as Dragon entered orbit, the Falcon 9's first stage arrived back on Earth, although it was unable to make the soft landing as planned. Meanwhile, Dragon was on orbit carrying a record 2,395 kilograms of cargo en route to the ISS. In what was almost routine, Dragon began its series of orbit-raising maneuvers to prepare for an arrival at the ISS on 12 January, with a planned capture time of 11:12 UTC. On 11 January, Dragon executed a burn to place it in a 346 by 352 kilometer orbit and, by the end of the day, the vehicle was just 10 kilometers below the outpost's orbit at a range of less than 1,000 kilometers. The following day, Samantha

Cristoforetti and Barry Wilmore prepared for the berthing of Dragon from inside the ISS Cupola. Capture occurred at 10:54 UTC, which was ahead of schedule. The rendezvous had been by the book and had run ahead for almost the entire timeline. As with SpX-4, Dragon was berthed (at 13:54 UTC) with the Harmony module and hatch opening was planned to occur the following day. After the hatch had been opened on 13 January, the crew went to work on cargo operations by first removing time-critical life sciences payloads, which included an experiment that studied regenerative mechanisms in flatworms (see sidebar for an explanation of this and other experiments). Also removed was the CATS payload but, despite the acronym, this had nothing to with felines. CATS was the Cloud Aerosol Transport System (see sidebar) that had been ferried up to the ISS bolted inside Dragon's trunk. As with previous payloads that had been transported in the trunk, the Canadarm 2 was required to retrieve CATS, but this time the Canadian robotic arm had a helping hand from the Japanese Remote Manipulator System (JRMS).

CATS

The CATS (Figure 5.15) is a remote sensing instrument that uses Light Detection And Ranging (LIDAR), Doppler LIDAR, and High Spectral Resolution LIDAR to measure clouds and aerosols to provide scientists with an idea of the impact of pollution on climate. It is hoped that the data generated by CATS will enable a hazard warning capability for events such as dust storms and wildfires. To do its job, CATS uses lasers operating at three wavelengths (355, 562, 1064 nm) and a telescope that measures the backscatter of light produced by aerosols in the atmosphere.

5.15 The Cloud Aerosol Transport System – CATS. This system will be used to measure atmospheric cloud profiles that will in turn help meteorologists better understand climate feedback processes. Credit: NASA

Fruit Fly Lab-01

During SpX-4 the life sciences experiment that garnered all the attention was the rodent study. On SpX-5 the experiment of interest was the fruit fly lab. Why fruit flies? Scientists can study fruit flies (*Drosophilia Melanogaster*) to see how microgravity affects the insects with special attention on illnesses since about three-quarters of human illness genes are very similar in the fruit fly genome. There's no need to euthanize fruit flies either because they have very short life spans. The Fruit Fly Lab-01 (Figure 5.16) studies fly under two conditions: one group is kept in a cassette that is under the influence of weightlessness while another group lives in another cassette which is subjected to artificial gravity.

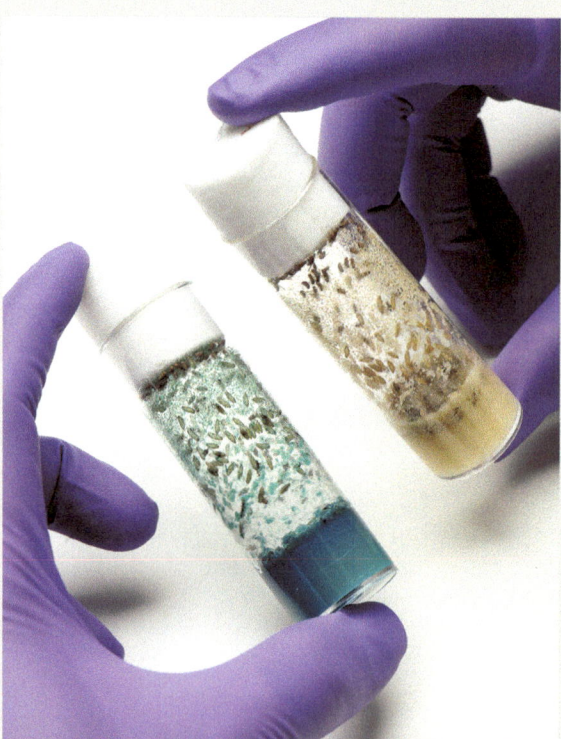

5.16 The Fruit Fly Lab. This system will study *Drosophilia melanogaster* – the common fruit fly. Why? Well, about three-quarters of human disease genes are analogous to the fruit fly genome. This lab will house more than 100 flies for up to 30 days. Credit: NASA

Flatworm Regeneration

In addition to rats, mice, and flies, ISS crews are also studying flatworms (Figure 5.17). The mechanism under study in the flatworms is tissue generation because these creatures have the ability to regenerate new cells by simply replacing them.

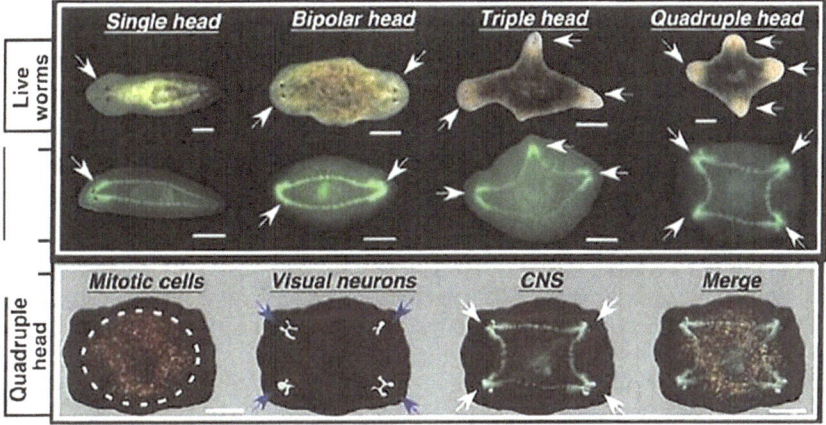

5.17 Flatworms. This experiment is all to do with regenerative medicine because flatworms happen to have extraordinary regenerative powers. By studying these creatures, scientists hope to be able to understand how microgravity affects healing. To do that, researchers will study regeneration patterns in these flatworms and monitor their tissue patterning and immunohistochemistry. Credit: NASA

Once the Canadarm-2 had positioned the CATS payload within a robot's arm's reach of the JRMS, the JRMS took over and grappled the CATS from the Dextre and the Japanese arm transported the CATS into position at the Exposed Facility Unit 3, where the payload was hooked up to the necessary data and power connections.

On 10 February, Dragon wrapped up its four-week stay at the ISS and was prepped for departure. Packed with 1,662 kilograms of downmass, the Canadarm-2 once again helped the vehicle on its way under the ever watchful eyes of Mission Controllers. Departing the ISS at 19:10 UTC, Dragon spent almost five hours in free flight before preparing for its de-orbit burn and closure of its GNC bay door, which occurred at 22:42 UTC. Dragon executed its de-orbit burn at 23:49 and splashed down off the California coast at 00:44 UTC on 11 February.

CRS6/SpX6

CRS6/SpX-6 was the seventh Dragon resupply mission and the sixth of 15 CRS flights that SpaceX has contracts to complete. In addition to ferrying 2,015 kilograms of cargo to the ISS, SpaceX was once again attempting to land the first-stage booster on the Autonomous Spaceport Drone Ship (ASDS). Launch was scheduled for 10 April 2015 at 20:33 UTC. On 8 April, Falcon 9 was rolled to the launch pad and placed in the launch position, after which the vehicle and Dragon were powered up and the launch team conducted a countdown operation. Following the routine static fire test and Launch Readiness Review, SpaceX confirmed that Falcon 9 was ready for launch (Table 5.7).

Unfortunately, due to a cluster of electrically charged clouds that entered the launch safety zone, the first launch attempt was scrubbed just three minutes before lift-off. But, the next day, on 14 April, the sixth commercial resupply mission was underway when the Falcon 9 blasted off at 20:10 UTC (Figure 5.19). Ten minutes after launch, Dragon was delivered into its planned orbit and began its by now customary chase of the ISS with a planned arrival on 17 April. The Falcon 9 performed flawlessly with the first stage, burning for just over two and a half minutes before separating from the second stage which burned for another seven minutes to boost Dragon into its orbit. With its job done, the first stage began its series of landing maneuvers to attempt a controlled return to the ASDS which was located 345 kilometers downrange from the launch site. After completing the boost-back firing, the first stage performed a three-engine burn and used its grid fins to fly to the ASDS. The first stage fired its center engine, deployed the landing legs, and descended to the deck of the ASDS, but the landing was a little squirrely and, although the first stage touched down, it tipped over due to excessive lateral velocity. Shortly after the failed landing, SpaceX posted video taken from a drone that had recorded the final seconds of the approach. The video showed the first stage arriving vertically but the lateral velocity was too quick and probably resulted in the overshoot of the platform despite the center engine gimbaling in an effort to correct the flight path. When the Falcon 9 booster impacted

Table 5.7 SpX-6.

Time	Event
T–00:03	Merlin 1D engine ignition
T–0	LIFT-OFF
T+00:18	Pitch and roll
T+01:10	Mach 1
T+01:24	Maximum dynamic pressure
T+02:38	Main engine cut-off
T+02:41	Stage separation
T+02:49	Stage 2 ignition
T+03:05	Stage 1 maneuver out of stage 2 plume and stage 1 reorientation
T+03:28	Dragon nose-cone jettison
T+04:00	Stage 1 apogee at 140 km
T+04:30	Stage 1 boost-back burn
T+06:30	Stage 1 grid fin deployment
T+06:45	Stage 1 re-entry burn (duration: 15 sec)
T+08:00	Stage 1 landing burn
T+08:20	Stage 1 landing leg deployment
T+08:35	Stage 1 landing
T+08:55	Flight termination system safing
T+09:36	Stage 2 cut-off
T+10:10	Dragon separation
T~12:00	Solar array deployment

5.19 Launch of CRS-6 flight. Credit: NASA

the deck, it created quite a debris field, with pressure bottles and other equipment being ejected overboard.

Meanwhile, on orbit, Dragon was performing as advertised, with the vehicle generating power and all its systems functioning nominally. Just over two hours after launch, Dragon

5.20 European Space Agency (ESA) astronaut Samantha Cristoforetti keeps an eye on Dragon from the Cupola. Credit: NASA

opened its GNC doors, exposing the grapple fixture and the navigation sensors required for the approach to the ISS, which was in an orbit of 396 by 404 kilometers. Using its HA and CE burns, Dragon closed in on the ISS and entered the 28-kilometer communications zone around the ISS five hours prior to capture on 17 April. Employing AGPS, Dragon made its burns to reach its holding position 2.5 kilometers below and behind the orbiting outpost, by which time operations between SpaceX and NASA Mission Control were tracking the vehicle to ensure it was in the right position for rendezvous. Dragon conducted its Approach Initiation Burn at 0.3 meters per second while positioned behind and below the ISS's R-bar. At 1,000 meters from the ISS, Samantha Cristoforreti (Figure 5.20) and Terry Virts monitored the approach from the Cupola using external cameras and telemetry data from Dragon. At the 350-meter point, Dragon flipped around to place the vehicle in the correct attitude to direct the Dracos in the right orientation for the posigrade maneuver in the event of an abort. Resuming its climb to the ISS at 09:20 UTC, Dragon employed the DragonEye sensors to ensure accurate navigation to the next checkpoint at 250 meters below the ISS. Following a 13-minute hold at 250 meters, Dragon resumed its approach at 09:27 UTC. As it passed the 200-meter checkpoint, the vehicle entered the KOS and Dragon gently pulsed its thrusters to maintain its position in the approach corridor. After taking 34 minutes to travel from 250 to 30 meters, Dragon was held for a systems check before being approved for final approach. Final approach took another 17 minutes, after which the

5.21   A welcome addition to the ISS cuisine, Lavazza's ISSpressor machine. Credit: NASA

vehicle entered free drift mode and the robotic arm grappled the vehicle under the control of Samantha Cristoforetti. Capture was announced at 10:55 UTC.

After leak checks were completed, the crew went about the task of outfitting the vestibule and hooking up Dragon to ISS power before focusing their attention on the cargo items (Appendix V). Among the most talked-about items was the capsule-based ISSpresso machine (Figure 5.21), which had been manufactured for the crew by Italian company Lavazza. Another notable payload item was the Rodent Research 2 experiment which was to investigate the degradation of the immune system in space. Also flown was the Fluid Shifts Study, which is perhaps one of the most complicated experiments ever flown on the ISS. The study will try to measure the volume of fluid shifting from the lower body to the upper body during the transition from gravity to spaceflight. Hopefully this information will provide scientists with data that will help them solve the problem of how fluid shifts affect the vision impairment problem currently afflicting many long-duration astronauts.

Following on from CRS6, CRS-7 launched on June 28, 2015, but it disintegrated 139 seconds into the flight shortly before first stage separation from the second stage. A month after the flight, SpaceX posted the following statement on their website:

On June 28, 2015, following a nominal liftoff, Falcon 9 experienced an overpressure event in the upper stage liquid oxygen tank approximately 139 seconds into flight, resulting in loss of mission. This summary represents an initial assessment, but further investigation may reveal more over time. Prior to the mishap, the first stage of the vehicle, including all nine Merlin 1D engines, operated nominally; the first stage actually continued to power through the overpressure event on the second stage for several seconds following the mishap. In addition, the Dragon spacecraft not only survived the second stage event, but also continued to communicate until the vehicle dropped below the horizon and out of range.

SpaceX has led the investigation efforts with oversight from the FAA and participation from NASA and the U.S. Air Force. Review of the flight data proved challenging both because of the volume of data—over 3,000 telemetry channels as well as video and physical debris—and because the key events happened very quickly.

From the first indication of an issue to loss of all telemetry was just 0.893 seconds. Over the last few weeks, engineering teams have spent thousands of hours going through the painstaking process of matching up data across rocket systems down to the millisecond to understand that final 0.893 seconds prior to loss of telemetry.

At this time, the investigation remains ongoing, as SpaceX and the investigation team continue analyzing significant amounts of data and conducting additional testing that must be completed in order to fully validate these conclusions. However, given the currently available data, we believe we have identified a potential cause.

Preliminary analysis suggests the overpressure event in the upper stage liquid oxygen tank was initiated by a flawed piece of support hardware (a "strut") inside the second stage. Several hundred struts fly on every Falcon 9 vehicle, with a cumulative flight history of several thousand. The strut that we believe failed was designed and material certified to handle 10,000 lbs of force, but failed at 2,000 lbs, a five-fold difference. Detailed close-out photos of stage construction show no visible flaws or damage of any kind.

In the case of the CRS-7 mission, it appears that one of these supporting pieces inside the second stage failed approximately 138 seconds into flight. The pressurization system itself was performing nominally, but with the failure of this strut, the helium system integrity was breached. This caused a high pressure event inside the second stage within less than one second and the stage was no longer able to maintain its structural integrity.

Despite the fact that these struts have been used on all previous Falcon 9 flights and are certified to withstand well beyond the expected loads during flight, SpaceX will no longer use these particular struts for flight applications. In addition, SpaceX will implement additional hardware quality audits throughout the vehicle to further ensure all parts received perform as expected per their certification documentation.

As noted above, these conclusions are preliminary. Our investigation is ongoing until we exonerate all other aspects of the vehicle, but at this time, we expect to return to flight this fall and fly all the customers we intended to fly in 2015 by end of year.

While the CRS-7 loss is regrettable, this review process invariably will, in the end, yield a safer and more reliable launch vehicle for all of our customers, including NASA, the United States Air Force, and commercial purchasers of launch services. Critically, the vehicle will be even safer as we begin to carry U.S. astronauts to the International Space Station in 2017.

6

NASA's Crewed Development Program

© Springer International Publishing Switzerland 2016
E. Seedhouse, *SpaceX's Dragon: America's Next Generation Spacecraft*,
Springer Praxis Books, DOI 10.1007/978-3-319-21515-0_6

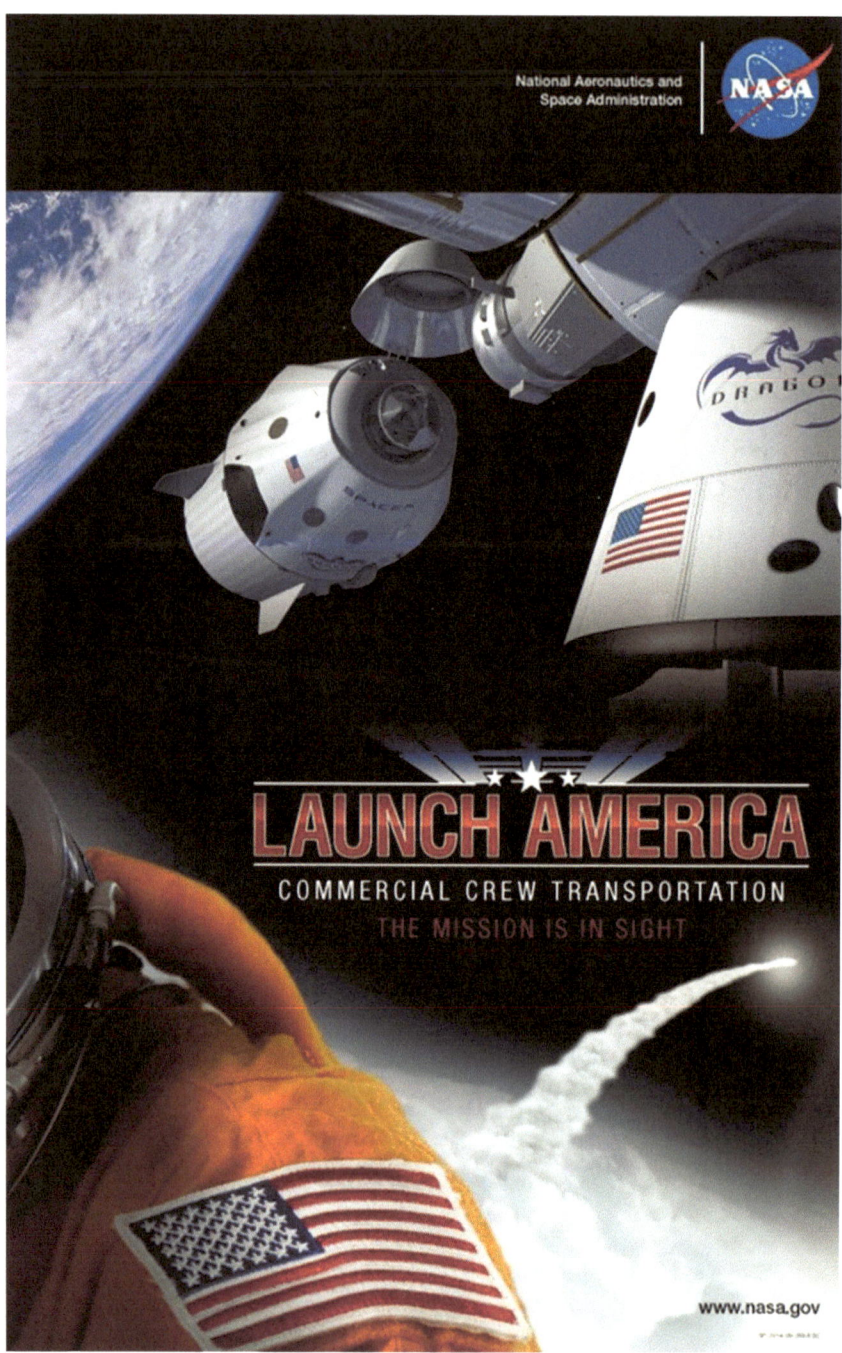

NASA's Commercial Crew Program. Credit: NASA

Before we move on to discuss Dragon Lab and the development of the manned Dragon variant – Dragon V2 – it is instructive to take a look at the other commercial crew transportation efforts. The commercial ball really got rolling in April 2011 when NASA awarded four Space Act Agreement (SAAs) in the second round of the Commercial Crew Development (CCDev2) initiative (see Chapter 2). The recipients of the agency's funding were SpaceX, Boeing, Blue Origin, and Sierra Nevada Corporation. With the money awarded to them by NASA, the companies were expected to advance crew transportation concepts, develop the design, and mature elements of the systems along a timeline that stretched to May 2012. While Dragon was by far the most well-known of the winning proposals, Sierra Nevada's Dream Chaser had also generated a lot of positive press thanks in part to the vehicle looking like a baby Shuttle (Figure 6.1). In fact, in the world of commercial manned spaceflight, Dream Chaser was something of a fan favorite.

DREAM CHASER

As you can see in Figure 6.1, Dream Chaser is designed to launch on top of an Atlas V, which places Sierra Nevada in the complete system provider category. With their US$80 million award, Sierra Nevada planned to develop and mature the Dream Chaser to free flight capability. Unlike Dragon, which splashes down in the ocean (although it will eventually perform rocket-assisted landings), the Dream Chaser will land on a conventional runway, just like NASA Langley's HL-20 spaceplane (Figure 6.2) on which the Dream Chaser is based. Thanks to the extensive HL-20 and spaceplane (BOR-4, X-24A, and HL-20) heritage available to Sierra Nevada, the company had a head start on development; for example, the HL-20 program alone generated 1,200 wind tunnel and simulation tests. The baby Shuttle also counts many astronauts as fans thanks to the vehicle's low-impact landing, low G-loading during re-entry, and no black zones during ascent. It also features a Return To Landing Site (RTLS) capability. Designed to carry up to seven astronauts, structural testing of Dream Chaser began in December 2010 at a lab at the University of Colorado. The engineering test article, which was built almost entirely of composites, was also the flight vehicle for the free flight test.

To prepare for the free flight test, Sierra Nevada performed a captive carry test in May 2012. Lifted by a Sikorsky Skycrane helicopter, a full-scale Dream Chaser without propulsion elements was put through a series of banks at 10,000 feet and at a speed of 90 knots near the Rocky Mountain Metropolitan airport in Colorado. The test was listed as Milestone 12 on the list of CCDev2 objectives as follows:

> Optional Milestone 12: ETA Captive Carry Flight Test Description: Conduct ETA captive carry flight test on carrier aircraft to characterize integrated vehicle performance. Success Criteria: Completion of ETA captive carry flight as outlined per the flight test plan.

This milestone was contingent on completion of Milestone 18 which was a readiness review of Milestone 12. After completing the captive carry test, Sierra Nevada went to work testing computational fluid dynamic (CFD) models in preparation for the free flight test (Optional Milestone 19). A month later, in June 2012, Dream Chaser passed Milestone 17,

6.1 Sierra Nevada's Dream Chaser. Credit: NASA

6.2   NASA's HL-20 spaceplane. Credit: NASA

which was the preliminary design review (PDR). The PDR is critical for any spacecraft and includes a review of all the key elements of the flight program, of not just Dream Chaser, but also the support ground systems. With the PDR checked off, the way was open for integrated system testing and an approach and landing test (ALT) schedule that would use the Dream Chaser Engineering Test Article (ETA).

"The successful completion of this full system PDR established that the Dream Chaser has a credible system design which is now approved to proceed towards integrated system testing."
John Curry, Director of Dream Chaser Systems Engineering, Integration, Test, and Operations

Working with NASA's Dryden Flight Test center at Edwards Air Force Base, Sierra Nevada now needed to find a carrier aircraft. Unlike the Shuttle Enterprise that used the Shuttle Carrier Aircraft (SCA) – a repurposed 747 – Sierra Nevada were looking for something smaller. Meanwhile, in the summer of 2012, Sierra Nevada announced that its preferred launch vehicle was the Atlas V in its 402 configuration. If all went to plan, Sierra Nevada hoped to launch Dream Chaser atop the 402 from the Atlas V LC-41 launch pad sometime in 2017, although this date would be contingent not only on Dream Chaser being complete, but also on United Launch Alliance (ULA) finishing work on man-rating the 402. The reason Sierra Nevada chose the 402 was because it was a known quantity and it was reliable. Another positive was the Emergency Detection System (EDS) that ULA was developing – a system that will be critical to the crew during the four-minute terminal count and through orbital insertion. In addition to the announcement of the 402, Sierra Nevada also revealed more details of how Dream Chaser would operate. Sierra Nevada planned a two-month turnaround of the vehicle. During this time, sections of the Thermal Protection System (TPS) would be replaced if needed and the docking adapter would be replaced.

In July 2012, Sierra Nevada completed yet another milestone by testing its vehicle's nose landing gear (NLG). In common with the Shuttle, Dream Chaser will touch down using its Main Landing Gear (MLG) but, when its nose is pitched forward, it will not use a traditional NLG for roll-out, instead relying on a skid strip. Sierra Nevada completed several tests of the MLG and skid strip and presented the results to NASA, who approved the tests. Following the tests of the landing gear, Sierra Nevada had passed 18 of their CCDev2 milestones and were in good shape heading into the ALT. Before that occurred, Sierra Nevada, along with SpaceX and Boeing, were informed they were winners in NASA's Commercial Crew Integrated Capability (CCiCap) competition. CCiCap was the sequel to CCDev2 and provided Sierra Nevada with more agency funding to develop Dream Chaser. The awards were announced in August 2012, at which time SpaceX was still considered the leader in the commercial crew competition, since they had already flown Dragon to the International Space Station (ISS). Although winning a smaller (US$212.5 million) award than aerospace behemoth Boeing (US$460 million), Sierra Nevada were not expected to develop Dream Chaser to the Critical Design Review (CDR) stage, but were still expected to fulfill several system baseline and integrated system safety reviews. And the pressure was on because it was possible that NASA might ultimately down-select just one company in future contracts. That decision would be based on the myriad pros and cons of the competing designs and could also potentially lead to a lifting body versus capsule tug-of-war.

By the end of 2012, Sierra Nevada were developing their flight-test program that would culminate with a mission to the ISS. Remember, 2012 was the first year that American astronauts had been without a manned spaceflight capability, being forced to hitch rides with the Russians. While Dream Chaser seemed to be winning the hearts and minds of the commercial manned spaceflight community, there was no guarantee it would survive the next funding round. In May 2013, the Dream Chaser ETA headed to Dryden where it was prepped for a series of drop tests. It was in Dryden that NASA Administrator Charlie Bolden (Figure 6.3) met with the Sierra Nevada team and had the opportunity to fly Dream Chaser using the spacecraft's simulator. Bolden, a veteran of four Shuttle missions, apparently had an easy time landing Dream Chaser at Edwards Air Force Base. In addition to the welcome attention from NASA's highest-ranking official, Dream Chaser also earned the praises of William Gerstenmaier, NASA's Associate Administrator for human exploration and operations, who had this to say about Sierra Nevada's efforts:

"Unique public-private partnerships like the one between NASA and Sierra Nevada Corporation are creating an industry capable of building the next generation of rockets and spacecraft that will carry U.S. astronauts to the scientific proving ground of low-Earth orbit. NASA centers around the country paved the way for 50 years of American human spaceflight, and they're actively working with our partners to test innovative commercial space systems that will continue to ensure American leadership in exploration and discovery."

William Gerstenmaier

6.3 Charlie Bolden addresses the employees of Sierra Nevada Corporation in May 2013. Credit: NASA

As Dream Chaser continued being prepared for its CCiCap testing, Sierra Nevada continued work on the vehicle's hybrid rocket motor[1] and on developing the Flight Test Article (FTA) which would be an upgraded version of the ETA. The FTA would be the second vehicle built by Sierra Nevada, but it wouldn't be the spacecraft that would ferry astronauts to orbit. For that, Sierra Nevada planned to build a third vehicle – the Orbital Vehicle (OV).

Commercial crew certification

In August 2013, while Sierra Nevada was testing its rocket motor, NASA announced its plans for commercial crew certification, which comprised three phases. Phase 1 was simply a procurement contract based on Federal Acquisition Regulations (FARs), while Phase 2, which was also a FAR-based contract, would be a contract to be awarded in the summer of 2014. This contract would include Design, Development, Test, and Evaluation (DDTE) requirements that would be needed to certify a crew transportation system. One of the requirements to be certified was performing at least one manned test flight to the ISS in the period between July 2014 and September 2017. NASA went on to explain that they hoped there would be more than one provider during this phase but also emphasized that they thought it unlikely the agency would continue funding three providers during this phase.

---

[1] This was tested in 2010 under the CCDev1 SAA. Incidentally, Sierra Nevada also manufactured the rocket motors for Virgin Galactic's SpaceShipTwo.

With a down-select on the horizon, Sierra Nevada knew there was pressure to demonstrate just how impressive Dream Chaser was, and one of the first ways to do that was the drop test. Unfortunately, Dream Chaser's first flight didn't go as planned when, on 26 October 2013, the ETA suffered a landing gear failure. Until the landing gear glitch, the flight had gone swimmingly, with all systems in the green. Dream Chaser, having been released from the Erickson Air-Crane helicopter, flew free for the first time, pulled up from her dive, and began her glide towards Edwards Air Force Base's Runway 22L. The baby orbiter flared and touched down on the centerline but a mechanical failure of the left landing gear resulted in a loss of control and Dream Chaser skidded off the runway. The damage to Dream Chaser was minimal and the key objective had been achieved: proving that the vehicle could fly. Sierra Nevada was happy with the test and announced after the flight that the company had already started work on their orbital version of Dream Chaser. In December 2013, despite Dream Chaser's landing gear gremlin, NASA confirmed the vehicle had passed the agency's CCDev2 milestone requirement. Buoyed by the news, Sierra Nevada pushed on to completing the next milestone of their CCiCap contract. Milestone 7, which was the Certification Plan Review (CPR) for the Dream Chaser, was completed in January 2014. It was perhaps the most important step because Milestone 7 confirmed Sierra Nevada's certification strategy and demonstrated to NASA that the spacecraft could comply with the functional and performance requirements of operating in orbit. Things were looking good, but the details of the looming down-select process were murky at best. One rumor was one company would lose out and the two remaining companies would be funded at full and half funding. Other rumors placed SpaceX and Sierra Nevada as the front runners. Who to believe? Sierra Nevada made it clear they were confident of being one of those selected for further funding. Steve Lindsay, Sierra Nevada's Dream Chaser program manager and five-time Shuttle pilot, had this to say:

"I had the privilege of piloting and commanding five Space Shuttle flights as a NASA astronaut. This included the last flight of Discovery which was processed, launched, and on March 9, 2011, made its final landing at the SLF after 39 flights and 148 million space miles. Mark, the entire SNC Dream Chaser team, and I look forward to seeing Dream Chaser continue this legacy from Discovery when it flies in 2016."

Steve Lindsay speaking at the Shuttle Landing Facility (SLF)

As the calendar counted down to the down-select, Sierra Nevada continued business as usual. In May 2014, shortly after Dragon splashed down from SpX-4, Dream Chaser passed Milestone 8 – the wind tunnel test, which involved testing the flight dynamic characteristics the vehicle will be subjected to during ascent and re-entry. Boeing was also keen to impress, showing the media the interior of their CST-100 capsule. But, of the three leading contenders, SpaceX was expected to be in a position to launch a crew first – a factor that would surely score well with those deciding where the funding dollars should go. While Dream Chaser was a firm favorite among commercial space fans, there was also the fact that NASA was siphoning off hundreds of millions of dollars to the Russian space agency for seats on board the aging Soyuz and the sooner America had a manned space capability the better.

In July 2014, Dream Chaser passed Milestone 9 on the CCiCap checklist. Milestone 9 was a test of Risk Reduction and Technology Readiness Level which demonstrated that all five major spacecraft systems[2] had advanced in accordance with CCiCap objectives – a requirement that was the culmination of more than 3,500 tests on the five systems. As the CCtCap decision date approached, the smart money was on Space X and Sierra Nevada to progress into the next phase of the Commercial Crew Program, so it was more than a little surprising that the winners of the pot of funding money were SpaceX and Boeing. How did Sierra Nevada miss out? While it was always very unlikely that the transition from CCiCap to CCtCap would have allowed all three companies to continue to receive agency funding, how could NASA deselect a company that had been such a front runner? It's difficult to say because the nuts and bolts of the selection process are classified. To many outsiders who had been following the story in the media, the decision made very little sense. After all, industry juggernaut Boeing had said they probably wouldn't be able to continue developing the CST-100 without agency funding – Boeing, whose operating cash flow (before pension contributions) was US$9.7 billion in 2013! In contrast, Sierra Nevada, whose total 2013 revenue amounted to US$2 billion, had always stated they would continue to develop Dream Chaser regardless of the CCtCap decision. Seeing US$4.2 billion being handed over to Boeing was hard to take (SpaceX received US$2.6 billion) and Sierra Nevada vowed to fight the decision, filing a protest with the General Accountability Office (GAO). It was the first legal challenge the company had ever filed.

As a result of missing out on a slice of the US$6.8 billion, Sierra Nevada had to lay off 9% of its Colorado workforce, although the company remained financially stable. In its protest, Sierra Nevada argued there were inconsistencies in the source selection process because Dream Chaser had achieved scores that were comparable to SpaceX and Boeing. As Sierra Nevada waited for the outcome of their protest, the company indicated it would continue developing Dream Chaser regardless of the outcome. In October 2014, at the American Society for Gravitational and Space Research (ASGSR) conference, it revealed the Dream Chaser for Science, or DC4Science. This Dream Chaser variant is designed to fly for short missions, providing customers with a vehicle on which to conduct science in the fields of biotechnology, life sciences, and material and fluid science. Another concept announced was the 75% Dream Chaser, which was a scaled-down variant of the original. Why 75%? A study had calculated this was the optimum size to be launched by Stratolaunch (Figure 6.4). While only being capable of carrying three crewmembers, the 75% variant could probably still dock with the ISS.

In January 2015, the US GAO turned down Sierra Nevada's protest, stating:

"In making its selection decision, NASA concluded that the proposals submitted by Boeing and SpaceX represented the best value to the government. Specifically, NASA recognized Boeing's higher price, but also considered Boeing's proposal to be the strongest of all three proposals in terms of technical approach, management approach, and past performance, and to offer the crew transportation system with

---

[2] Crew Systems, Environmental Control and Life Support Systems, Structures (1,500 tests), Thermal Control and Thermal Protection Systems (350 tests).

© Sierra Nevada Corporation

6.4   The Stratolaunch system with a Dream Chaser attached. Credit: Sierra Nevada

most utility and highest value to the government. NASA also recognized several favorable features in the Sierra Nevada and SpaceX proposals, but ultimately concluded that SpaceX's lower price made it a better value than the proposal submitted by Sierra Nevada.

"In making its selection decision, NASA concluded that the proposals submitted by Boeing and SpaceX represented the best value to the government. Specifically, NASA recognized Boeing's higher price, but also considered Boeing's proposal to be the strongest of all three proposals in terms of technical approach, management approach, and past performance, and to offer the crew transportation system with most utility and highest value to the government. NASA also recognized several favorable features in the Sierra Nevada and SpaceX proposals, but ultimately concluded that SpaceX's lower price made it a better value than the proposal submitted by Sierra Nevada."

A couple of weeks later, NASA weighed in on why they had passed on Dream Chaser, citing complexity of design and uncertainty of when the vehicle might be ready to fly to the ISS. The agency acknowledged that Boeing's vehicle was a more expensive design[3]

------

[3]   SpaceX's price in its Crew Dragon proposal was US$1.75 billion, Boeing's price was US$3.01 billion, and Sierra Nevada's was US$2.55 billion, according to a GAO statement released on 5 January that upheld NASA's decision on the commercial crew contracts. Those prices do not include the extras bundled into the final contract values, such as fully fledged crew rotation flights once the CST-100 and Crew Dragon vehicles are certified by NASA.

but also that the CST-100 was the strongest in terms of mission suitability. Boeing also received high marks for technical and management approach. Sierra Nevada was rated the lowest for level of maturity, with the agency also noting that Dream Chaser's propulsion system had not been finalized. Some in the commercial space industry were quick to point the finger at politics, noting that Boeing simply had more power in Congress. There may be some truth to that argument. After all, Congress tried for years to kill commercial space to the benefit of Boeing and Lockheed. By the time the CCtCap decision came along, NASA's commercial crew program was underfunded by hundreds of millions of dollars and deselecting Boeing wouldn't have helped their case. So, the only way to persuade Congress to continue funding commercial crew programs was to award a contract to Boeing: it was basically NASA acknowledging that they couldn't beat Boeing but that they could perhaps meet them halfway. But what about the real assessment? After all, you didn't need to have an engineering degree to know that SpaceX was way ahead of Boeing and Sierra Nevada in the design of their vehicle. So why Boeing ahead of Sierra Nevada? Speed was probably the answer. NASA needed a manned spaceflight capability to the ISS yesterday and they figured that Boeing would probably be able to deliver the goods faster than Sierra Nevada, given the funding climate at the time. In the long term, Dream Chaser is way ahead of the Gemini-era capsule designs of Boeing and SpaceX, but Boeing already knows how NASA wants things done, so why not go ahead with a less ambitious but known quantity? That's what NASA did. It's a shame because, by 2015, many of us probably expected our means of ferrying astronauts to the ISS to be little more advanced than stuffing crews into a tin can and giving them a parachute and an airbag.

BOEING CST-100

Perhaps the least well-known commercial vehicle in the CCDev2 competition was Boeing's CST-100 (Figure 6.5) for which the company received US$92 million in the first round of funding. Designed to carry seven crew and pressurized cargo to low Earth orbit (LEO), the CST-100 can be launched on a selection of launch vehicles (Figure 6.6) and can be reused up to 10 times. At the time of the CCDev2 award, Boeing reckoned they would be able to provide services by 2015, which also happened to be the target year announced by the other award winners.

Boeing released 25 milestones at the CCDev2 presentation. Some of the more notable milestones included a landing air bag drop demonstration, wind tunnel tests, and a parachute drop test. As the test program approached the PDR, Boeing planned to test the service module propellant tanks and the launch vehicle EDS, the latter objective matching the work being performed on the human rating of the Atlas V. The Atlas V, which was selected by Boeing as their launcher in August 2011, is a two-stage rocket powered by the Russian-built RD-180 engine. At the time of the announcement, the Atlas V had logged 26 unmanned launches since its first flight in 2002 and had achieved a 100% mission success.

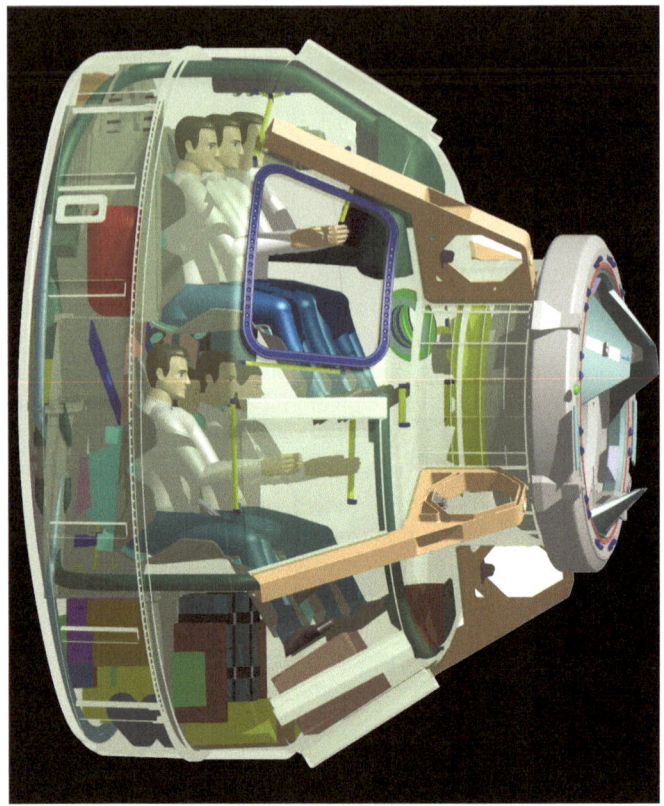

6.5   Boeing's CST-100 vehicle. Credit: NASA

Boeing CST-100 Spacecraft

The CST-100 draws on Boeing's experience with the Shuttle and ISS programs, although it has no Orion heritage. The vehicle will use NASA's docking system and Boeing's own lightweight ablator heat shield. With a diameter of 4.56 meters and a length of 5.03 meters, the CST-100 is designed to spend up to 210 days docked with the ISS.

In April 2012, Boeing conducted the first in a series of drop tests to assess the effectiveness of the CST-100's airbags. In the first test, the CST-100 test vehicle was dropped from an Erickson Sky Crane from an altitude of more than 3,000 meters above the Delmar Dry Lake Bed in Nevada. The vehicle's parachutes duly did their job, slowing the descent, after which the airbags inflated to ensure a soft ground landing. The second test (Figure 6.7), which was conducted the following month from an altitude of 4,200 meters, was

6.6   The Atlas V carrying the Juno spacecraft. Credit: NASA

performed with support from Bigelow Aerospace who may use the CST-100 to ferry commercial customers to its inflatable habitats one day.

After the success of the drop tests, Boeing looked forward to August 2012 when NASA was due to announce CCiCap funding. To no one's surprise, Boeing was one of the winners, pocketing US$460 million (SpaceX received US$440 million and Sierra Nevada received US$212.5 million). One of the next key tests on the milestone list was the Mission Control interface test, which took place in August 2013. This test was essentially an interconnectivity test that confirmed Boeing's ability to send and receive data from Mission Control to its avionics software integration facility. The interface test was followed by a test of the CST-100's thrusters. This test was performed in September 2013 at White Sands

6.7   Boeing's CST-100 drop test. Credit: NASA

Space Harbor in New Mexico. The tests, which were conducted with Aerojet Rocketdyne, put the CST-100's orbital maneuvering and attitude system (OMAS) through its paces by firing the vehicle's thrusters in a vacuum chamber. The successful test meant Boeing had completed its ninth milestone and was now on track to complete all its CCiCap objectives by mid-2014.

In April 2014, the media had the opportunity to check out the interior of the CST-100 when Boeing unveiled a mock-up of their new vehicle (Figure 6.8). While many had complained about the less-than-futuristic capsule concept, there was little to grumble about inside the vehicle. "Minimalistic" and "lean" were two words that came to mind when peering inside the very smart interior. No cluttered panels of switches in this spacecraft. Unlike previous capsule designs, the CST-100 had a look that bore more of a resemblance to a Gulfstream jet than a spacecraft. For those who work in the aerospace industry, the design shouldn't have come as a surprise because Boeing has routinely been winning awards for innovative interiors for aircraft for almost as long as there have been aircraft.

"We are moving into a truly commercial space market and we have to consider our potential customers – beyond NASA – and what they need in a future commercial spacecraft interior."

Chris Ferguson, ex-Space Shuttle Atlantis *commander and*
Boeing's Director of Crew and Mission Operations

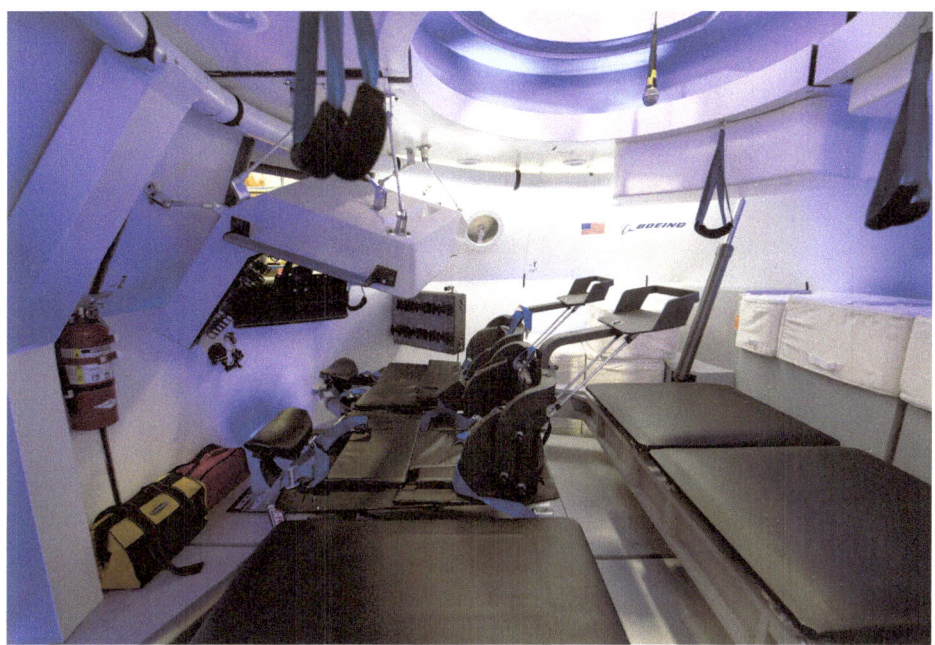

6.8 Interior of the CST-100. Credit: NASA

As you can see in Figure 6.8, the signature color is blue and that blue lighting is based on Boeing's Sky Lighting that is very similar to the technology sported by the company's Dreamliner. Unlike the Shuttle, the CST-100 is no place for bulky manuals, thanks to elegant tablet technology and crew interfaces.

In August 2014, after the company had passed NASA's CDR, Boeing announced that its CST-100 vehicle was on track to launch astronauts in 2017. This meant that the agency was satisfied that all the systems and sub-systems that comprise the CST-100 worked together. The CDR, which also included a hazard analysis of the spacecraft, opened the way for NASA and Boeing to set the design of the vehicle and approve the ground segment CDR. This milestone was passed in December 2014 following a three-week review of Boeing's plans for building and operating the CST-100. The ground segment CDR included an assessment of Boeing's plans to build the vehicle, how they were going to get ready to fly, how the CST-100 would be placed on the launch vehicle, and how the vehicle would operate once it was in orbit. Such a comprehensive review required the input of an assortment of engineers, safety and performance experts, and, of course, astronauts. With the ground segment CDR out of the way, Boeing could now press ahead with plans to actually build the vehicle and ULA could go ahead designing the launch vehicle adapter and the business of man-rating their Atlas V.

In December 2014, Boeing announced it had developed a cargo variant of the CST-100. The cargo variant would be a scaled-down version of the man-rated vehicle, since it wouldn't need a launch abort system or the life-support system. Given the competition

Boeing faced from companies such as SpaceX and Sierra Nevada, perhaps the announcement wasn't too surprising. Meanwhile, progress continued on developing the CST-100 and the infrastructure to support it. Construction of the crew access tower was underway at the Atlas V launch pad and work continued to man-rate the launcher in preparation for its 80th flight, which was planned as the first crewed flight test (the 74th Atlas V flight launched in January 2015). The Orbiter Processing Facility (OPF High Bay 3), the future home of the CST-100, was also being repurposed in anticipation of receiving Boeing's hardware. By March 2015, flight software for the CST-100 had already been developed and Boeing had a flight simulator operational. The flight manifest showed a pad abort test scheduled for February 2017 and an orbital unmanned flight test two months later. The crewed flight test, to be piloted by a Boeing and a NASA astronaut, will be flown in July 2017.

BLUE ORIGIN

If it's one thing Blue Origin is famous for, it is the quiet way it goes about its business. Most people became aware of this secret squirrel enterprise in 2006 when the company, headed by Amazon.com founder Jeff Bezos, lifted the veil on their rocket project by flying a demonstration vehicle at its West Texas launch facility. But, except for a few photos of the vertical launch and landing rocket, Blue Origin kept details of its Goddard vehicle shrouded in secrecy. In 2010, Blue Origin were a little more open when the company revealed it had been awarded US$3.7 million by NASA to develop an astronaut escape system and to build a prototype ground test vehicle. The escape system (Figure 6.9) was described as a "pusher" system that placed the rockets on the base of the capsule instead of the tower, as was the

6.9 Blue Origin performs a pad escape test. Credit: NASA

case in the Apollo era. In addition to working on the escape system, Blue Origin was also continuing to develop its Goddard, which was the first vehicle in the company's New Shepard program. A suborbital vehicle, New Shepard was being designed to fly from Blue Origin's spaceport in Texas to an altitude of 120 kilometers. But Blue Origin also had its sights on orbital spaceflight, which is why it wasn't surprising when it was awarded US$22 million in NASA's CCDev2 awards in August 2011. Little was revealed about Blue Origin's plans for their biconic capsule, although it was being designed to carry seven astronauts and be capable of being berthed at the ISS for 210 days in a lifeboat role.

In May 2012, Blue Origin's vehicle completed a series of more than 180 wind tunnel tests at Lockheed Martin's High Speed Wind Tunnel Facility. The tests confirmed the vehicle's aerodynamic capabilities during re-entry and also its ability to alter its flight path. In addition to the wind tunnel tests, Blue Origin was conducting tests of its BE-3, a liquid-oxygen, liquid-hydrogen engine which would be used in a pad abort test planned for later that year. Before the pad abort test, Blue Origin conducted a flight test of their second test vehicle. Unfortunately, the test, which took place in August 2012, didn't go too well when an in-flight failure occurred at more than 14,000 meters while the vertical-take-off, vertical-landing (VTVL) vehicle was traveling at Mach 1.2. Better news followed two months later when the company conducted a successful pusher escape pad test at its West Texas launch facility. In the test, the pusher escape motor launched a full-scale crew vehicle to an altitude of more than 700 meters before descending under canopy to a soft landing. For those who might fly a suborbital trip in the vehicle, the test will provide peace of mind because it demonstrated full-envelope crew escape – a feature that is lacking in the SpaceShipTwo suborbital vehicle. The test also placed Blue Origin within striking distance of completing its Systems Requirement Review (SRR), although achieving that wouldn't translate into more money for the company because NASA had already made their decision in August to continue funding only SpaceX, Boeing, and Sierra Nevada. But Blue Origin continued developing their system under an unfunded NASA agreement – one that was extended on 31 October 2014 as part of their existing SAA. While the agreement does not provide funding, the agency provides technical guidance to Blue Origin as the company sought to fulfill its milestones using internal funding. In the same month as Blue Origin extended their SAA, ULA announced that it had signed an agreement with Jeff Bezos's company to develop a liquid-oxygen/liquefied natural gas engine. This announcement came after lengthy deliberations in Washington about the liability of relying on Russian-built RD-180 engines on the Atlas V launcher. The US Air Force (USAF), not wanting to rely on Russian engines any longer, had determined to replace the RD-180 as soon as possible. The replacement? Blue Origin's BE-4 engine. Predictably, Energomash, manufacturer of the RD-180, wasn't happy because the RD-180 is only used on the Atlas V. At the time of the announcement, most people were aware of the BE-3 but unaware of the BE-4, which had already been in development for more than two years. Engine testing is planned for 2016. Assuming the BE-4 is certified, ULA will be able to bring down launch costs to a level where it will be able to compete with SpaceX. At today's prices, a ride on the man-rated version of the Dragon will cost US$22.75 million whereas a ride on the Boeing CST-100 will cost more than US$35 million. But, if the BE-4 engine is realized, Boeing's cost may come down. Not only that, but Dragon may have more competition from Blue Origin because having the BE-4 available will mean the company has the capability to launch its space vehicle into orbit.

7

DragonLab

DragonLab. Credit: SpaceX

DragonLab first hit the news at the end of 2008 when SpaceX held an invitation-only meeting to publicize its free-flying version of Dragon. At the time of the meeting, SpaceX planned its first DragonLab mission in 2010 followed by a second in 2011, although no details were divulged about the customer. In common with Dragon, DragonLab is designed

© Springer International Publishing Switzerland 2016
E. Seedhouse, *SpaceX's Dragon: America's Next Generation Spacecraft*,
Springer Praxis Books, DOI 10.1007/978-3-319-21515-0_7

to be launched by a Falcon 9 and is capable of carrying 6,000 kilograms of upmass and 3,000 kilograms of downmass. Missions could last anything from a week to two years. DragonLab is good news for researchers because it promises regular commercial access to space and it can also return experiments. The only platform that was capable of doing this until recently was the Shuttle, but DragonLab (Appendix II) offers a few advantages over the Orbiter because it is not a crewed vehicle, which means payload safety requirements are lower and integration processes are faster. This all adds up to more frequent flight opportunities and faster turnaround for payloads. It also means the lessons learned from one flight can be quickly applied to a follow-on mission.

DRAGONLAB AS A MICROGRAVITY RESEARCH PLATFORM

DragonLab's (Figure 7.1) payload services are practically the same as Dragon, the details (Table 7.1 is included as a reminder) of which were described in Chapter 3, so the focus of this chapter is how DragonLab may be used for microgravity research. As you can see in Table 7.2, there are limited decreased gravity opportunities available for researchers and many of these options have limitations, the most common of which is the length of time in true microgravity. Take parabolic flight (Figure 7.2) as an example. This is achieved when an aircraft climbs steeply before pitching downward in a series of parabolas. Depending on the skill of the pilot, each maneuver will result in about 22–24 seconds of microgravity of 10^{-3} g. It's a great platform, and it's fairly inexpensive (about US\$4,500 per flight), but the length of microgravity is less than half a minute.

High-altitude balloon drops aren't much better when it comes to microgravity time. This platform offers a reasonably long period of free fall of up to 60 seconds during which

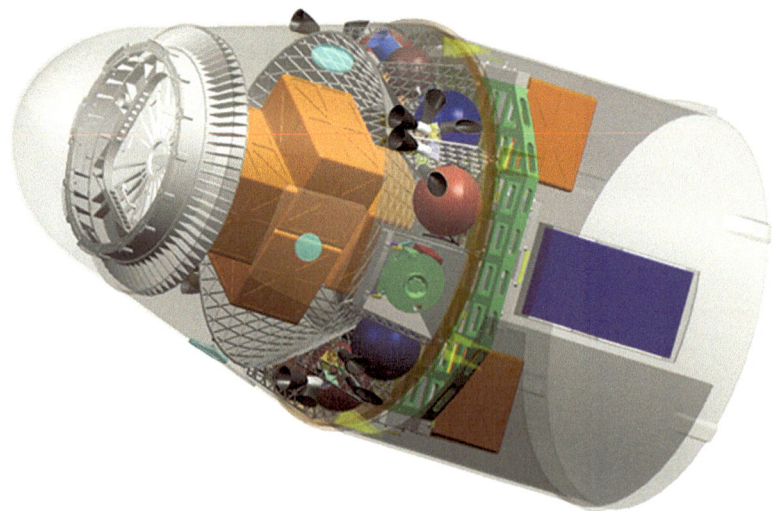

7.1 DragonLab payload configuration. Credit: SpaceX

Table 7.1 DragonLab.

Payload services	
Mechanical	*Thermal and environmental*

• Specific mounting locations and environments are mission-unique
• Pressure vessel interior (pressurized, recoverable)
  – 10 m³ payload volume
  – Lab temp, pressure, and relative humidity
  – Typically Middeck Locker accommodations
  – Other mounting arrangements available
• Sensor bay (unpressurized, recoverable)
  – Approx 0.1 m³ volume
  – Hatch opens after orbit insertion
  – Electrical pass-throughs into pressure vessel
• Trunk (unpressurized, non-recoverable)
  – 14 m³ payload volume
  – Optional trunk extension for a total of up to 4.3 m length, payload volume 34 m³
	• Internal temperature: 10–46°C
	• Internal humidity: 25–75% relative humidity
	• Internal pressure: 13.9–14.9 psia
	• Cleanliness: Visibly Clean–Sensitive (SN-C-0005)
	• Pressurized: convective or cold plate
	• Unpressurized: cold plate if required
	• Payload random vibration environment:
	  – Pressurized: 2.4 g (<100 lbm)
	• Unpressurized: 2.9 g
	Telemetry and command
	• Payload RS – 422 serial I/O 1553 and Ethernet interfaces (all locations)
	• IP-addressable payload standard service
	• Command uplink: 300 kbps
	• Telemetry/data downlink: 300 Mbps (higher rates available)

Power • 28 VDC and 120 VDC • Up to 1,500–2,000 W average: up to 4,000 W peak	

Ref. NASA SSP 57000 and www.spacex.com.

Table 7.2 Microgravity platforms.

Platform	Microgravity level	Experiment duration	Year	Sample return
Drop tower	10^{-2} to 10^{-5} g	Up to 10 sec	Current	Yes
Parabolic flight	10^{-2} to 10^{-3} g	Up to 20 sec	1959 →	Yes
Balloon drop	10^{-2} to 10^{-3} g	Up to 60 sec	Current	Parachute
Sounding rockets	10^{-3} to 10^{-5} g	3 to 6 minutes	1950's→	Parachute
Human-tended suborbital	10^{-2} to 10^{-5} g	3–6 min	2010→	Yes
Shuttle	10^{-2} to 10^{-5} g	Up to 17 days	1981–2010	Shuttle
EURECA	10^{-4} to 10^{-5} g	1 year	1992–93	Shuttle
Salyut 1–7	10^{-3} to 10^{-5} g	Up to 8 years	1971–86	Soyuz (limited)
Skylab	10^{-3} to 10^{-4} g	Up to 2 years	1973–74	Apollo (limited)
MIR	10^{-3} to 10^{-5} g	Up to 4 years	1986–2000	Soyuz (limited)
Bion/Foton	10^{-3} to 10^{-5} g	12–18 days	1973–92, 1985 →	Yes
ISS	10^{-3} to 10^{-6} g	12 days years	1998→	Shuttle – 2010
DragonLab	10^{-3} to 10^{-7} g	Hours to years	2010→	Dragon

7.2 Parabolic flight. Credit: ESA

a useful microgravity level of 10^{-3} g is achieved until atmospheric drag becomes a problem (when terminal velocity is reached gravity is once again 1 g).

Another option for microgravity researchers (Table 7.2) are sounding rockets (Figure 7.3). NASA has been sending these vehicles into space for more than 40 years as part of its Sounding Rocket Program and the European Space Agency (ESA) has a similar

7.3 NASA testing a suborbital sounding rocket: the Talos-Terrier Oriole. Credit: NASA

program. These rockets offer microgravity for up to 12 minutes, the missions are relatively inexpensive, and payloads can be developed quickly. Twelve minutes is a good chunk of time in microgravity, but what if you need longer? Well, then you have to find an orbital platform and, with the retirement of the Shuttle, that means either the International Space Station (ISS) or Foton (Figure 7.4).

7.4 Foton-M. These satellites are produced by TsSKB Progress and are operated by Roscosmos. Standing 7.73 m tall with a diameter of 2.7 m, the Foton-M can carry about 600 kg inside and 250 kg externally. Fotons have been used to carry all sorts of scientific payloads, ranging from radiation studies to housing a family of geckos. Credit: NASA

The Russian Foton spacecraft was in the news in July 2014 when the Russian space agency lost contact with their Foton M4 vehicle, which was carrying a group of geckos together with an assortment of Russian and German experiments. Although the vehicle was designed for lengthy autonomous operation, the failure of two-way communications meant that Russian controllers couldn't send any commands to the spacecraft. Fortunately, after a week of trouble-shooting, Mission Controllers established contact and the vehicle landed safely. But that's the Russian option, which you can take advantage of incidentally if you submit through ESA. To fly your payload on the ISS, you can submit your idea through NASA's Solicitation and Proposal Integrated Review and Evaluation System (NSPIRES) website: a review usually takes about seven months or so. Or you can take the DragonLab route, but what science are we talking about here?

DRAGONLAB SCIENCE

The primary fields of microgravity research include biology and life sciences, materials physics, fundamental physics, biotechnology, pharmaceuticals, and fluid physics. A detailed description of each of these categories is beyond the scope of this book but it is instructive to understand why the absence of gravity is an important element in experiments conducted in these fields because it underscores the significance of DragonLab will have in the future.

We'll start with life sciences. Experiments submitted in this field have received increased funding over the past few years partly due to current human space exploration goals. Most life sciences experiments share common features in terms of what capabilities are required: almost all require return to Earth, most require arming shortly before launch and access to the samples following landing – all capabilities that DragonLab will be able to accommodate. One example of a typical experiment that might be flown on DragonLab is in the field of plant physiology. With all the talk of sending astronauts on long-duration missions to asteroids and eventually to Mars, there has been a renewed focus on closed life-support systems that include growing food for the journey. One subject that life-support engineers are particularly interested in is *gravitropism* which is a term used to describe the ability of plants to align themselves with Earth's gravity. In space, the gravitational field is practically non-existent which means plants don't grow as they would on Earth. Some plants are affected by the absence of gravity than others – more gravitropically sensitive – and research is needed to determine which plant species grow the best in microgravity (Figure 7.5). In fact, as I'm writing this, astronauts on board the ISS are conducting plant studies on this very subject: in February 2015, a set of samples in JAXA's

(courtesy of Professor Takayuki Hoson of Osaka City University)

7.5 An example of circumnutation. Credit: NASA

Plant Circumnutation[1] and its Dependence on the Gravity Response experiment were being monitored by astronaut Terry Virts. The goal of the study is to determine whether microgravity has an effect on the circumnutation of rice. Why grow plants in spirals? Well, some plants may grow very tight spirals and some may grow in large curves – obviously with space at a premium, it is desirable to grow plants that take up the least space and that is what the JAXA study aims to find out.

Plant physiologists are also interested in the effects of radiation on plants because some plants are more radiation-resistant than others. DragonLab will provide an ideal candidate environment to provide answers to these questions. Another life sciences experiment that may find its way onto DragonLab is one related to the bone and muscle atrophy astronauts suffer while on orbit. The problem has been studied for years but the mechanisms that cause astronauts to lose bone density and cause their muscles to waste away are only partly understood. While DragonLab is an unmanned vehicle, it is possible animal experiments could be flown that would help answer these questions.

Another major microgravity field of research is pharmaceuticals and biotechnology. Much of the microgravity research conducted in this field is aimed at developing more effective drugs and treatment protocols, and one of the ways these goals can be realized is by increasing the understanding of transmembrane and intracellular transfer mechanisms. Another popular field is protein crystal growth because it has been shown that crystals grown in microgravity can be grown with a purity that is unachievable on Earth. This is helpful to the biotech industry because, by analyzing crystals grown in microgravity, it is possible to precisely determine the function of the many proteins used in the body. Take the respiratory syncytial virus (RSV), for instance. RSV is a deadly infectious disease that kills 4,000 infants every year. The study of the structure of the recombinant antibody that combats RSV has been conducted in microgravity and that research has helped scientists design more effective therapeutic strategies to deal with RSV. Diabetes research (Figure 7.6) is another area of research that could find its way onto DragonLab. By improving crystals in human insulin, diabetes treatment will be more effective: this goal could be achieved using a microgravity-based bioreactor to produce commercial proteins for that purpose. Another application is artificial organ transplant. One of the problems with artificial transplantation is rejection and one of the reasons the body rejects an organ is because cells grown on Earth are not three-dimensional due to the effects of gravity: once again, a bioreactor could solve that problem by growing commercial proteins that are very close to the structure of those in the body.

Next is materials physics. The most popular research in this field is focused on growing zeolite crystals. Zeolites (Figure 7.7) are very tough crystals that appear to have a honeycomb structure. Unlike a sponge that must be squeezed to release any fluid it might contain, zeolites release whatever they're holding when heated or when subjected to pressure. Since zeolites can soak up petroleum and still remain rock solid, they have proven very popular in the oil and gas industry and it isn't surprising that practically all the world's petroleum is produced using zeolites. But not all zeolites are equal. Zeolites that are near

---

[1]   Circumnutation describes a plant growth in which the plant bends into a spiral.

7.6　Crystals grown in microgravity. On the left are microgravity-grown crystals of recombinant human insulin, which are larger and have greater optical clarity than those grown terrestrially. Credit: NASA

7.7　Zeolite crystals. These are synthetic and were grown at the Center for Advanced Microgravity Materials Processing. Credit: NASA

perfect make the process of producing petroleum more efficient than zeolites that have defects and imperfections. And, since the US wants to reduce its dependence on foreign oil, there has been a big push to grow better zeolite crystals, which is part of the reason that the Zeolite Crystal Growth Furnace Unit (ZCG-FU) was flown on the ISS.

Materials physics is also interested in the transition from fluid to solid states. For example, in microgravity, it is possible to create alloys and polymers that cannot be manufactured on the ground, partly due to the fact that convection in materials is reduced when there is no gravity. By studying fluid flows in materials in microgravity, it has been

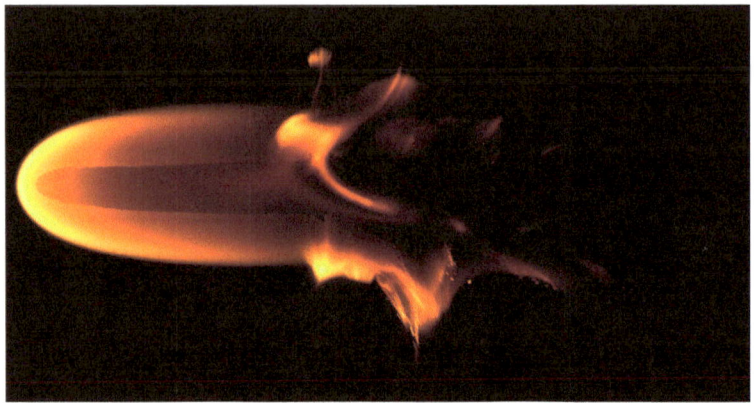

7.8 A flame in microgravity. Credit: ESA

possible to develop more efficient processing techniques which in turn has resulted in the production of stronger materials. Fluid physics on the other hand is more concerned with the study of heat and energy flows which can be studied better without the influence of convective forces. By studying foams, emulsions, and granular materials in microgravity, fluid physicists gain valuable insight into the way these materials can be processed. For example, analysis of surface tension-driven flows is much easier in microgravity, and study of this phenomenon has implications for techniques such as welding and semiconductor crystal growth.

Combustion physics is another major field of microgravity-based research that is perfectly suited for DragonLab. Materials don't burn (Figure 7.8) the same way in microgravity as they do on Earth and this has an impact on the materials used in spacecraft and how the life-support systems of spacecraft are designed. Other mechanisms that fall under the umbrella of combustion physics are gas diffusion, heat transfer, air flow, and pressure profiles in materials that are alight. The direct benefits of this research to those working on orbit is the development of more efficient ways to fight fires and the most effective ways to extinguish fire. Conducting combustion physics experiments on board the ISS is liable to make any mission manager a little twitchy but, for DragonLab, the safety requirements will be a lot more relaxed than they are on board the orbiting outpost.

In addition to fluid physics, combustion physics, and materials physics, there is also fundamental physics, which is interested in the states of solids, liquids, and gases, and the forces that affect them. One of the fundamental physics research platforms destined for the ISS is NASA's Cold Atom Laboratory (CAL) which will study ultra-cold atoms and ultra-cold quantum gases.

DRAGONLAB VERSATILITY

So that's the science that could be performed on DragonLab, but just how versatile is this vehicle as a microgravity platform? We'll start with the payload accommodations, which comprise the pressure vessel, the trunk, and the sensor bay. Combined, the three accommodations offer plenty of room for experiments and, even if some of these payloads happen to be on the heavy side, mass is unlikely to be a limitation because the Falcon 9 has plenty of lift capacity. This is important for researchers because many missions are mass-limited and not volume-limited. Also, having to reduce mass late in the development phase is not only a frustrating exercise, but it also usually results in a reduction in performance. Payload power shouldn't be much of an issue for most payloads, although power will be determined by orbit. For example, when DragonLab is flying a mid-inclination orbit, the payload power available will be between 1,500 and 2,000 watts whereas in certain sun-synchronous orbits, as much as four kilowatts may be available. What does this mean for the various science payloads discussed? Biological science experiments need up to 100 watts for the duration of an experiment that may last for a few hours, days, or weeks: with up 200 watts available, this means several such experiments could be carried on one mission. If researchers want to communicate with their payload, DragonLab offers uplink and downlink telemetry and command services: since the payloads are IP-addressable, scientists should be able to simply log on to access their payloads when the DragonLab makes a ground pass. Another DragonLab quality that will be appealing to scientists is the option to load payloads – particularly biological science and biotech payloads – as late as nine hours before launch and accessing the payload as soon as six hours after landing. Microgravity quality is also important to many customers, and DragonLab delivers on this condition too. For those dedicated microgravity users, DragonLab can easily achieve 10^{-6} g.

OTHER APPLICATIONS

DragonLab isn't limited to science and flying payloads. It could be used to deploy satellites, free-flying spacecraft, and research on the effects of the space environment. It could also be used as a platform to study the effects of orbital debris or the effects on materials caused by radiation. And it could be used to further Elon Musk's dream of traveling to Mars. How? It's all down to gravity, or a lack of it in this case. For as long as there have been astronauts, researchers have struggled to find a way to slow the insidious process of muscle atrophy and bone loss that occurs during long-duration missions. Science has had some success in creating countermeasures but a three-year trip to Mars is currently beyond the capability of the human body to endure. So what to do? Well, there are some who suggest scientists are approaching the problem from the wrong end: rather than trying to adapt the body to lack of gravity, why not just create gravity? Those who have been around a little longer than I have will remember the Gemini 11 and 12 missions that attached tethers to an Agena docking target before spinning the spacecraft end over end. While gravity levels were low, the Gemini–Agena configuration did generate some gravity. Unfortunately, the enthusiasm for artificial gravity dissipated after Gemini and there was little work on the subject until it was somewhat resurrected by Kent Joosten at NASA's Exploration Analysis and Integration Office in 2002 (Figure 7.9).

The designs suggested by Joosten are summarized in Table 7.3. The basic design was a 56-meter-radius configuration that spun at four revolutions per minute, creating 1 g. While the design was a breath of fresh air for those struggling with the mission concepts designed

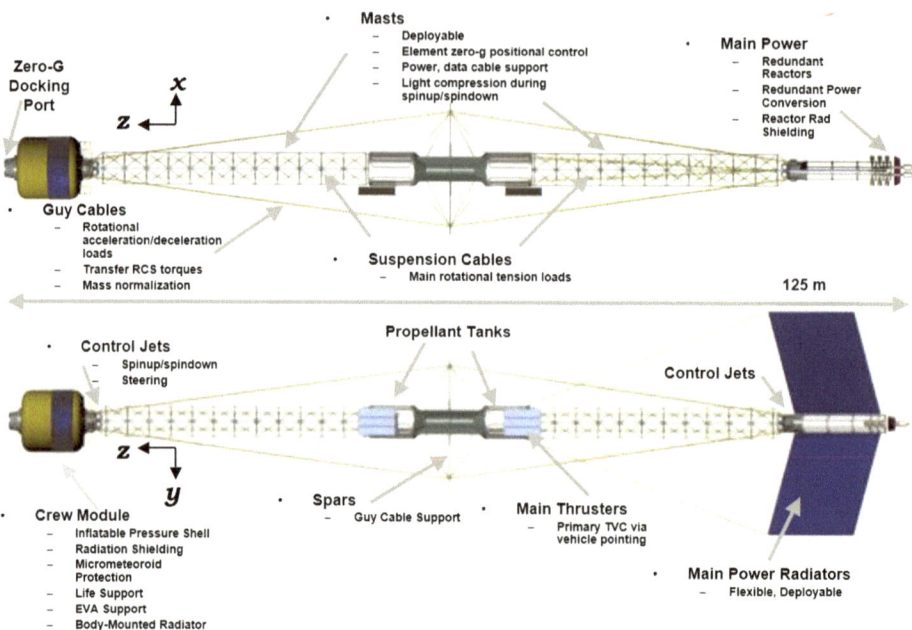

7.9 An artificial-gravity concept integrated into the design on a spacecraft. Credit: NASA

Table 7.3 Artificial-gravity concepts.

Concept	Features	Potential advantages	Potential challenges
Fire Baton	Hab counterweighted by reactor/power conversion systems Entire vehicle rotates Vehicle pointing provides majority of thrust vector control (TVC)	No rotating joints, power connections, fluid connections Power conversion systems operate in g-"field"	Vehicle angular momentum must be continuously vectored for TVC Thermal radiators in g-"field" Crew ingress/egress
Ox Cart	Hab counterweighted by reactor/power conversion systems Thrusters, despun, gimballed for TVC	Thrust vectoring decoupled from rotational angular momentum Power conversion systems operate in g-"field"	Megawatt-level power, prop transfer across rotating joints Potential cyclical loading of rotating joints Thermal radiators in g-"field" Crew ingress/egress
Beanie Cap	Split habitation volumes for counterweights Reactor/power conversion systems, thrusters in zero-g Thrusters gimballed for TVC	Thrust vectoring decoupled from rotational angular momentum Thermal radiators in zero-g	Inefficiencies in duplicating habitation systems, crew transfer between them Potential cyclical loading of rotating joints Power conversion systems operate in zero-g Kilowatt level power transmission across rotating joints

Adapted from Artificial Gravity for Human Exploration Missions NEXT Status Report July 16, 2002, B. Kent Joosten.

to ferry astronauts to Mars safely, there were very few data about hypo-g as a countermeasure and not that much about adaptation times to hypo-g. Nor was there much information about tether dynamics, tether design, or optimal spin-up methods. But this is where DragonLab could help.

Although the problems of human adaptation to hypo-g can't be answered by using DragonLab, there are still plenty of questions that must be answered before utilizing artificial gravity on a long-duration mission. For example, one of the most important challenges to resolve is determining the optimal spin-up method. Then there is the problem of tether dynamics and precession in particular. Precession is a problem because it can affect navigation, so there needs to be some way of dampening the effects of this. The tether design must also be tested because little is known about tether materials (except for a few missions flown from the Shuttle many years ago) and how strong they need to be to resist micrometeorite strikes. Another black hole of knowledge is how tethered systems respond to velocity changes and how attitude changes may affect power, communications, and velocity: for example, to what degree does an attitude change affect velocity? And, on the subject of communications, how will a spinning spacecraft maintain data rates? Finally, when the crew finally arrive at the Red Planet, they will need to disengage from the upper stage and use the capsule to land. How will this be done?

In addition to the lack-of-gravity problem, Mars mission planners must contend with the radiation mission-killer (Figure 7.10). While the radiation environment in low Earth orbit (LEO) is well characterized, the radiation beyond LEO is a very different animal and researchers know precious little about how it might affect a Mars-bound crew. Much of what is known was revealed during the Curiosity mission, which carried a Radiation Assessment Detector (RAD) instrument. During its journey to Mars, the toaster-sized

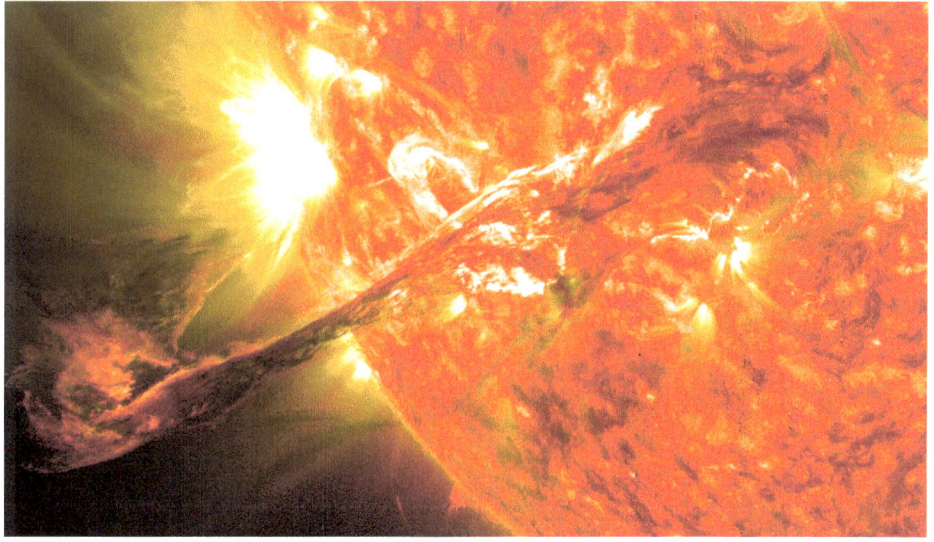

7.10 A solar flare. Credit: NASA

RAD measured all high-energy space radiation and the news wasn't encouraging because the average dose was approximately 300 mSv, or about 15 times the annual limit for someone working a nuclear power plant. An additional problem faced by Mars-bound astronauts is the notoriously fickle space weather, particularly solar events such as flares and coronal mass ejections. So how much shielding is required to protect astronauts from these events and how accurate is solar weather forecasting? DragonLab can help and here's how. A series of Mars-specific missions could be developed to answer the radiation and reduced-gravity questions. The first such mission could fly animals to measure responses to a partial-gravity environment and a microgravity environment. This could be achieved by creating a scaled-down configuration of Joosten's spinning spacecraft. A second artificial-gravity mission could conceivably be created for crew, although this would require some modifications inside DragonLab. This mission would dock with the ISS, take on board a couple of crew, and repeat the tasks of the first mission, after which the crew would return to the ISS. Follow-on missions could test various tether configurations and assess radiation levels in higher altitudes.

When will DragonLab fly? Initially, the first DragonLab mission was scheduled for 2010, but there have been several slips since then. As the manifest stands in mid-2015, the DragonLab Mission 1 is due to launch in 2016 sometime after SpaceX's 12th commercial flight to the ISS. DragonLab Mission 2 should now fly sometime in 2018 after the Iridium flight #7. By that time, there may be manned suborbital platforms such as the Lynx that may make a small dent in the microgravity research business, but DragonLab's versatility and orbital capability should prove a godsend to those engaged in microgravity research. Even today with the ISS up there, microgravity research is far from routine, but DragonLab could change that and, if it does, many fields of science and industry stand to benefit. It may also play a key part in determining how astronauts can get to Mars.

8

Preparing for Crew: Dragon V2

Artist's concept of the manned Dragon V2 docking with the International Space Station (ISS). Credit: SpaceX

In parallel with preparing Dragons for their commercial cargo missions, SpaceX was also busy developing its manned variant – the V2. In 2013, NASA had outlined its strategy for enabling the certification of commercial crew systems with a goal of restoring domestic manned launch capability by 2017. Phase 1 of this certification strategy was the Certification Products Contract (CRC), which was awarded to SpaceX, along with Boeing and Sierra Nevada in 2012. The contract was based on Federal Acquisition Regulations (FAR) that required the companies to adhere to specific standards such as hazards analysis, verification, and validation. The next phase of the contract went into action in the summer of 2014 with the awarding of funding to SpaceX and Boeing and the de-selection of Sierra Nevada. This phase of the contract outlined the Design, Development, Test, and Evaluation

© Springer International Publishing Switzerland 2016
E. Seedhouse, *SpaceX's Dragon: America's Next Generation Spacecraft*,
Springer Praxis Books, DOI 10.1007/978-3-319-21515-0_8

(DDTE) milestones that were required for NASA to certify a man-rated vehicle. Under the CCtCap certification, SpaceX was required to check off certain DDTE tasks, one of which included one manned test flight to the International Space Station (ISS) between July 2014 and September 2017. Once the first man-rated mission was checked off, NASA required at least two post-certification missions (PCMs) to be flown, although the maximum number of PCMs for each Boeing and SpaceX was six. These PCMs could be flown until 31 December 2020. Another important item encompassed special studies that Boeing and SpaceX would be required to perform to reduce risk in their man-rated vehicle, and a fourth item was complying with the cargo requirement of the manned vehicle, which was defined as a minimum of 100 kilograms of stowage volume, or an additional 100 kilograms for a seat that is not taken by a crewmember.

One of the items that could impact the timeline of certification flights and PCMs is the docking system that must be used on each spacecraft. One of NASA's requirements was that SpaceX and Boeing be able to dock their spacecraft with the ISS, with one option being to use a NASA Docking System (NDS) unit (Figure 8.1). As part of the award to SpaceX and Boeing, NASA made available four NDS units for no charge. The first unit was to be made available in February 2016, although, if either company didn't want to use the NDS, there was also the option to build their own, as long as it was compatible with NASA requirements as stated in the SSP 50808 document.

8.1　NASA's docking system. Credit: NASA

In addition to the docking system, NASA also had ensure the commercial vehicles had search-and-rescue (SAR) capability during ascent and re-entry, and that vehicles could execute pad aborts and emergency landings. A related item to the SAR and emergency capabilities was the issue of third-party liability that was required for any activity that was not covered by the Federal Aviation Administration (FAA) licensing. While FAA licensing was not required for the certification (test) flights, the licensing was a requirement for the PCMs which meant that commercial providers were required to have third-party liability insurance up to a maximum of US$500 million.

The CCtCap requirements also identified other items that could impact PCMs such as flying commercial passengers. For example, Bigelow Aerospace has sovereign agreements with a number of countries to fly commercial astronauts to his BEAMs (Bigelow Expandable Activity Module). These passengers could fly on either the V2 or Boeing CST-100, because Bigelow has agreements with both companies. But what impact would that have on PCMs? According to NASA, there is nothing preventing SpaceX or Boeing from manifesting a commercial passenger on a PCM as long as the company providing the flight adjusts the price in kind for NASA. Given that understanding, it is conceivable that commercial passengers could be bound for Bigelow's BEAMs (Figure 8.2) as early as 2018.

8.2 Bigelow's BEAM. Credit: NASA

DRAGON V2

> "You'll be able to land anywhere on Earth with the accuracy of a helicopter, which is something a modern spaceship should be able to do."
>
> *Elon Musk at the unveiling of Dragon V2*

The public had its first opportunity to see one version of the commercial future of American spacecraft in May 2014 when Elon Musk unveiled the crewed version of Dragon: Dragon V2 (Figure 8.3) – also known as Dragon 2. Before describing the nuts and bolts of the spacecraft, it is instructive to highlight the reason the V2 came to be. When the US lost its manned spaceflight capability in 2011 with the final flight of the *Atlantis*, NASA underwent a major recalibration, transitioning its goals to beyond Earth orbit (BEO) and paying Roscosmos exorbitant amounts of money to fly American crews to the ISS. At the same time, outsourcing became the name of the game and part of NASA's budget was directed to the Commercial Crew Program to develop commercial vehicles such as the V2. The irony that was not lost on most manned spaceflight observers was that the extra money NASA was saving from not having to operate the Shuttle was now being siphoned off to Russia: hundreds of millions of dollars every year. And, since the US has

8.3 Dragon V2. Credit: NASA

Table 8.1 Dragon V2 at a glance.

Crew	7		
Launch Vehicle	Falcon 9 v1.1	Payload	To the International Space Station (ISS): 3,310 kg From the ISS: 2,500 kg
Height	6.1 m	Endurance	1 week to 2 years
Diameter	3.7 m	Re-entry	3.5 s
Sidewall angle	15°	Thrusters	8 SuperDracos in four pods
Dry mass	4,200 kg	Propellant	NTO/MMH
Features			
Reuse	Up to 10 times	Docking	Autonomous capability
Landing	Propulsive and parachute	Heat shield	PICA-X 3rd generation

had no crew capability since 2011 and will not have that capability until at least 2017, that means the Russians have enjoyed some great paydays.[1]

One of the most striking features of the V2 is the propulsive landing technology that aligns with Musk's reusability aspirations. While the V2 is developed, Space X will test propulsive landings using a test vehicle dubbed DragonFly. The key to the propulsive landing capability, which we'll get to shortly, is the SuperDraco rocket engine, which is built in pairs and which will be used for launch aborts as well as landing crews on the ground. Unlike cargo Dragons, the V2 (Table 8.1) won't be grappled and berthed by the robotic arm thanks to the aforementioned docking system that commercial vehicles must be fitted with. Another difference between the cargo Dragon and the V2 is the heat shield, which will be an upgraded version of the PICA-X that is flown on the baseline Dragon. Although the V2 is designed to land propulsively, the vehicle has the redundancy of a parachute system and an extra layer of redundancy thanks to the capability of landing safely even if two thrusters fail.

The interior of the V2 looks very much like what you would expect a spaceship of the 2010s to look like. Luxury couches and elegant glass control panels. A clean, graceful, and very stylish design – and roomy too. The V2 will have seats for seven crewmembers. The astronauts will be seated on two tiers of seats with four on the upper tier and three on the lower (Figure 8.4). The commander and pilot will sit in the center of the upper tier with access to the control panel, while the crewmembers sitting either side of the commander and pilot will have the best seats because they will be seated next to large windows. As the V2 approaches the ISS, it will open the nose cover to ready the NDS for docking, which can be executed either manually or automatically. On its return from space, the V2 will utilize the hypergolic propellants that are stored in carbon tanks located around the edge of the V2 and pressurized helium that is stored in similar tanks. And this is when the SuperDracos will go to work.

---

[1]   To give you some idea of the cost, consider the price of just one seat deal. For example, in April 2013, NASA signed a deal with Russia for six seats at a cost of $70.7 million per seat for a total cost of $424 million. The previous contract had priced seats at $62.7 million per seat.

8.4   Dragon V2 interior. Credit: NASA

SuperDraco

The SuperDraco (Figure 8.5 and Table 8.2) was developed using the experience SpaceX gained developing the Draco engines that the cargo Dragon uses to execute orbital maneuvers when arriving and departing the ISS. The first test of the SuperDraco, which took place in 2012 at SpaceX's Rocket Development Facility in Texas, was an impressive demonstration of what the engine could do, not only sustaining its 71,000 newtons (16,000 pounds) of thrust (more than 200 times the thrust of the Draco) for the full duration of the test, but also executing the throttling capability that will allow the V2 to make precision maneuvers on orbit.

Eight of the engines that you see in Figure 8.5 will be installed on the V2 – an arrangement that gives the man-rated Dragon more than 500 kilonewtons of thrust if needed. And that thrust will be available immediately, because SpaceX designed the SuperDraco to reach its full thrust in just one-tenth of a second once the ignition switch is armed. For those interested in the finer points of the engine's design, the SuperDraco's chamber was created using a three-dimensional printing process, and that chamber is regeneratively cooled by routing the fuel through a jacket around the combustion chamber. If things go pear-shaped during launch, the SuperDracos would fire for five seconds, pulling the vehicle away from the launch vehicle. And, because those engines are a part of Dragon, SpaceX's launch abort system is available through the entire flight profile. No jettisoning of towers in this set-up. And if one of those SuperDracos were to fail? Not to worry because the vehicle can still perform a launch abort.

8.5   SpaceX's SuperDraco. Credit: NASA

Table 8.2   SuperDraco by the numbers.

Nozzle exit diameter	20 cm
Propellant hypergolic	NTO/MMH
Exhaust velocity	2,300 m/sec
Mass flow rate	321 kg/sec
Thrust:	16,000 lbf
	(71,196 N)
	(7, 257 kg-force)
	Note: maximum thrust: 16,400 lbf
Combined thrust (8 SuperDracos)	120,000 lbf (534 kN or 54,431 kg-force)

Since 2012, SpaceX had embarked upon an aggressive test schedule, performing full-duration burns, full-thrust burns, deep-throttle demonstrations, and tests in a variety of off-nominal conditions. Already by the end of 2012, the SuperDraco had 58 test firings under its belt and, by May 2014, SpaceX announced the engine had completed its qualification testing. On 6th May 2015 SpaceX completed the first of its pad abort tests, which are required as part of NASA's CCiCap. After half a second of vertical flight, Dragon's SuperDraco engines powered the capsule from 0 to160 kilometers per hour in just 1.2 seconds, reaching a maximum velocity of almost 600 kilometers per hour. The second pad abort will be a high-altitude test launching atop a Falcon 9 that will occur at 73 seconds into the flight. Why 73 seconds? Well, that's when maximum dynamic pressure occurs and SpaceX wanted to create the toughest test possible.

In addition to the abort tests, SpaceX are busy preparing the DragonFly program to demonstrate just how versatile Dragon V2 is. The DragonFly is a low-altitude vehicle that SpaceX will use as a test platform to demonstrate how the SuperDraco engines can be used to maneuver Dragon V2. As part of 30 DragonFly test flights, SpaceX will conduct at least two propulsive assist landings with parachutes and another two flights without parachutes. Of the other 26 flights, there will be 18 full propulsive tests and eight propulsive assist tests with an engine burn time of 25 seconds. Assuming DragonFly hits all the test objectives, SpaceX will have a propulsive landing capability, which means Dragon V2 will be able to land almost anywhere, although the preferred landing location will be Cape Canaveral. Landing at the Cape means SpaceX will cut the cost and time of hauling Dragon across country and also reduce the time and money refurbishing their spacecraft. In the optimal scenario envisaged by SpaceX, Dragon lands at the Cape, is refueled, checked, and sent on its way again.

"I haven't seen anything proposed that would match the Dragon Version 2 on a technology basis. This is really taking things to a new level."

Elon Musk at the unveiling of Dragon V2, May 2014

The key to Mars

SpaceX's CEO is absolutely right and, of all the cutting technology on Dragon V2, the most revolutionary is the propulsive landing capability – not just because it looks cool and not just because this is how many people think this is how spacecraft should land, but because it is the key element enabling a manned mission to Mars, and we all know that is the ultimate goal of SpaceX. There are many Mars advocates today who say we have the technology to ferry humans to Mars safely. We don't, and we're a long way from making a manned mission to the Red Planet a reality, but at least SpaceX with its propulsive landing Dragon V2 is taking a huge step in the right direction. To date, more than 60% of all missions to Mars have failed because it's very, very difficult to land a payload on the surface, even a small one. People assume that, since astronauts have landed on the Moon, it shouldn't be much of a stretch to do the same on Mars. The Apollo lander weighed about 10 tonnes and, even if all the provisions for staying on the surface of Mars were sent ahead of time, 10 tonnes would be the absolute minimum weight of a future Mars lander. But, in 2015, the very best engineers on the planet can't figure out how to land one tonne on the surface! Why? Before we get to that, let's consider the problem. The entry, descent, and landing (EDL) headache is also called the Supersonic Transition Problem (STP). In simple terms, it means that, with current EDL capability, a large vehicle plunging through the tenuous Martian atmosphere has about 90 seconds to decelerate from Mach 5 to Mach 1, flip over from being a spacecraft to being a lander, open the chutes to decelerate some more, and then fire those thrusters to navigate to the landing site before touching down. To do that is not only fiendishly difficult, but impossible using the technology we have today, despite all the claims of Mars One or other Mars proponents. But why can't we use airbags (Figure 8.6) say the Mars crowd? After all, they've been used pretty successfully on other missions. Yes they have, but those were unmanned missions with small payloads. And those airbag landings don't come without a punch: using the same principle for a

8.6   Airbag testing at Langley's Impact Research Facility. Credit: NASA

manned landing would subject the occupants to deceleration forces exceeding 20 Gs or more. Robots may be able to sustain such forces. Humans can't.

What about the Sky Crane then? This was the system (Figure 8.7) used to deliver the Mars Science Laboratory (MSL) to the surface. Since the rover weighed 775 kilograms, NASA had to devise a whole new landing architecture because the rover was too big for airbags. What they came up with was a ballistic entry combined with heat shield followed by a parachute, thrusters, and finally a crane-like configuration that lowered the rover on a tether so the rover landed on its wheels. Very clever. Ingenious in fact, but the system can't be scaled up for manned vehicles.

For vehicles returning to Earth, our planet's thick atmosphere causes spacecraft to slow to below Mach 1 at about 20 kilometers' altitude. The rest of the descent is dealt with by parachutes (Dragon) or drag and lift (Shuttle). But the Red Planet's atmosphere is just 1% as thick as Earth's, which means that the atmosphere is so thin that it is equivalent to the air density 35 kilometers above the surface on Earth. It doesn't matter what combination of existing EDL systems you use on Mars, the only thing that will happen if you try to land a manned vehicle is that you will be a crater on the surface. Supersonic parachutes won't work because they would need to be 100 meters in diameter and a chute that big probably can't be opened safely at supersonic speeds. If not supersonic parachutes, how about supersonic decelerators? Perhaps. The technology we're talking about is the Hypercone (Figure 8.8), which looks a little like an over-sized doughnut. This "doughnut" would encircle the spacecraft and inflate at about 10 kilometers above the surface while the

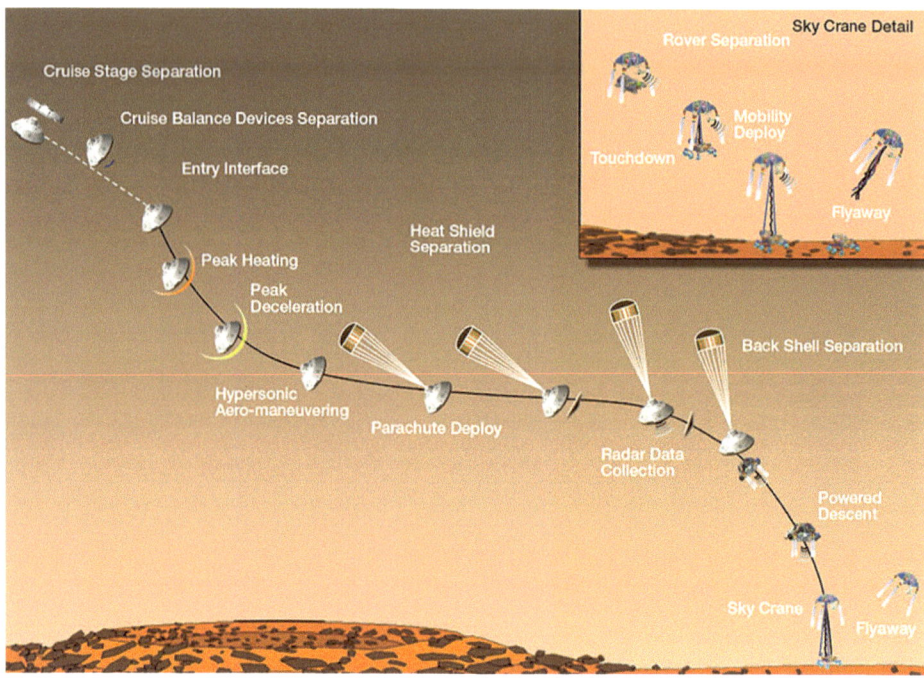

8.7 Sky Crane operations. Credit: NASA

8.8 Hypercone. Credit: Vorticity/NASA

vehicle was at Mach 4 or 5. Acting like an anchor, the Hypercone, which would be about 30 meters in diameter when inflated, would decelerate the vehicle to Mach 1, at which point subsonic parachutes would take over. The promising part of the Hypercone is that it can be scaled for vehicles as heavy as 20 or 30 tonnes. The not-so-good part is that we know that flexible structures of that size are very difficult to control. Plus, you only have a very small window to open those parachutes. And, assuming you can open the parachutes in time, you would still need to use thrusters to land.

So what to do about landing on Mars? The answer? Supersonic retro propulsion (SRP). You see, the key to landing a crew safely on the surface of Mars is decelerating to subsonic speed at least eight kilometers above the surface. Only by doing this do you give the crew enough time to make landing decisions and actually land the vehicle. And the only way to achieve this is to use a lot of deceleration. To their credit, NASA has begun work on this problem through their Propulsive Descent Technologies (PDT) project, but the spotlight is on SpaceX's propulsive landing system because they have proven that they can get things moving quickly. And the key to SpaceX achieving their propulsive landing goal is the SuperDraco. With no turbo-pump and no ignition, combined with that dual redundancy, it seems there is very little that can go wrong. Having said that, the system will have to be proven and then some before engineers commit human lives to engines, especially when they have just survived the heat of re-entry. Even with a tried and tested SRP system, there are still a long chain of events that have to go like clockwork for a vehicle to land safely. With unmanned vehicles, mission planners had the luxury of risking a little but, in a manned mission, everything must work flawlessly not just most of the time, but every time.

Launch abort system

Another revolutionary element of the SuperDraco is its use as Dragon's launch abort system (LAS). In most spacecraft systems, the LAS has usually comprised a tower attached to the top of the crew vehicle. In the event of an emergency, the tower would pull the spacecraft away from the launch vehicle and, if everything went well, the tower would be jettisoned sometime during the ascent. If you watch Hollywood movies, you may remember Tom Hanks as Jim Lovell reaching forward to jettison the tower during the flight of *Apollo 13*. Other LAS options have been utilized over the years, including ejection seats during the Gemini program and the first four flights of the Shuttle program.

In the history of manned spaceflight, the use of a LAS has been a very, very rare event. But it has happened. In September 1983, cosmonauts Vladimir Titov and Gennady Strekalov were anticipating the launch of their Soyuz 7K-ST No. 16L (also known as Soyuz T-10a or T-10-1) mission to the Salyut space station when the launch vehicle caught fire. A spaceflight veteran, Strekalov knew something was wrong and told Titov to tighten his harness. Moments later, the Soyuz escape tower blasted the capsule several kilometers from the pad while subjecting the crew to 20 Gs.

LAS was on the drawing board again following the *Columbia* accident when a crew safety system was an essential element of the Vision for Space Exploration (VSE) venture. The VSE never happened, but before it was cancelled, NASA had conducted several trade studies to determine which system worked best. And that work didn't go to waste because the design was eventually base-lined into the Orion design (Figure 8.9). That design comprises a tractor tower design that should save the crew from an on-pad abort, a mid-altitude emergency, and a high-altitude emergency. The system was tested using a boilerplate Orion in 2010.

And SpaceX's LAS? Despite having very little experience in the business of designing a LAS, the system SpaceX has come up with is arguably the most elegant and safe system

8.9 Orion's launch abort system (LAS). Credit: NASA

not only because the system is integrated into Dragon, but also because it can be used repeatedly. Traditionally, escape systems have used solid propellant because the stuff ignites quickly and full thrust is attained rapidly, but the SuperDracos are liquid engines. That isn't a problem though, because tests have shown that the engines provide full thrust within 100 milliseconds of the ignition command. That's fast. Plus, another advantage the SuperDracos have is their throttleability. Not only that, but the SuperDraco LAS has an extra layer of redundancy built in by virtue of being able to recover the vehicle even if one of the abort engines is lost. And then there's that integration. No need to jettison anything from Dragon. And that integration confers a greater abort envelope.

DRAGON V2 OPERATIONS

The cost of a seat on board Dragon V2? About US$22 million, which is about US$50 million less than what the Russians are charging. And, according to SpaceX, that US$22 million figure is based on a low flight rate. If Bigelow's customers start lining up for flights to the BEAMs, then that flight rate goes up and the seat costs come down, perhaps to as low as the single digit millions. That scenario would fit perfectly with Elon Musk's vision of how spaceflight should be run. Musk has never made any secret of the fact that his long-term vision is of thousands of spaceflights every year together with bases on the Moon and Mars. Musk's motivation for starting SpaceX in the first place was the lack of progress beyond Apollo. He remembers reading about the pronouncements that NASA would establish a base on Mars in the mid-1980s and that dramatic improvements in rocket technology would revolutionize manned spaceflight, but of course none of that ever happened. Perhaps thousands of flights every year is thinking a little too far down the line, but Musk reckons that hundreds of flights per year can be achieved within the next 15–20 years. And two key elements in that busy launch manifest are the SuperDraco engine and the Dragon V2. While the first version of the V2 will be refurbished after every 10 flights, the plan is to increase the number of flights without any major overhaul. SpaceX aim to eventually increase the lifetime of their Dragon V2 to as many as 100 flights, thanks in part to reduced ablation of the PICA-X heat shield. It should be noted at this point that the Dragon V2 could never have been developed without Dragon 1.0. SpaceX learned an awful lot designing, developing, and then flying Dragon. They learned about engine development through the Draco engine, they learned how to maneuver in orbit, how to de-orbit, and how to execute a precise re-entry path. The plan for Dragon 1.0 is for it to be gradually phased out but, for the next few years, SpaceX plan to develop the cargo Dragon and Dragon V2 in parallel. Incidentally, for those who wonder about the cost, SpaceX estimate that the development of Dragon through to Dragon V2 by 2015 was about US$400–500 million and that the final cost to get the Dragon V2 man-rated will probably be close to US$1 billion. When you consider that NASA had already spent US$5 billion by the time of the Orion test flight in December 2014, that's quite a bargain. Of course, between 70% and 80% of that cost is thanks to NASA funding, but US$1 billion for a man-rated vehicle nowadays is still a steal. And the seat cost? Well, that depends. The figure of US$22 million per seat was mentioned earlier, but that number needs some perspective. Remember, NASA awarded US$4.2 billion to Boeing and US$2.6 billion to SpaceX, but these awards are variable and cover between two and six operational flights. In NASA parlance, the operational flights are designated as PCMs. At a minimum, under the terms of the awards, Boeing and SpaceX must execute three flights in total, one of these being a test and the other two being a PCM, or operational flight. Either Boeing or SpaceX could be awarded up to seven flights, which would include one test and six PCMs to be executed before 2020. If Boeing were to fly seven flights, its total award would be US$4.2 billion, whereas, if SpaceX performs seven flights, its award will be US$2.6 billion. Under the Commercial Crew contract, SpaceX has always said that its flight price would be about US$140 million, which would equal about US$20 million per seat if Dragon V2 flew at full capacity. Now multiply US$140 million by seven and you get US$980 million, which means that the cost of Dragon's development would be US$2.6 billion minus US$980 million, or US$1.62 billion.

And Boeing? Well, the aerospace juggernaut hasn't been as accommodating when it comes to discussing the cost per seat on their CST-100 but, not so long ago, a seat price of US$36.75 million was quoted for a flight to a BEAM. Since the CST-100 can also fly seven astronauts, the flight cost of the CST-100 would equal US$257 million, or US$117 million more expensive than a SpaceX flight. And, if you continue the arithmetic, seven flights using Boeing's CST-100 would cost US$1.79 billion, which means that the development cost of the CST-100 would be about US$2.4 billion, or almost US$800 million more than Dragon!

BIGELOW

"This is going to be a fabulous machine. SpaceX deserves all the credit in the world. This is really a fork in the road for space exploration."

Robert Bigelow, founder of Bigelow Aerospace, speaking with NBC
News after testing Dragon's seats at the unveiling in May 2014

While SpaceX's first customers on board Dragon will almost certainly be government astronauts, for Musk to make money, he needs to find other clients because there is only one destination for NASA, CSA, and European Space Agency (ESA) astronauts, and that is the ISS. Fortunately, there is another pioneer in the commercial spaceflight arena and his name is Robert T. Bigelow (Figure 8.10). Bigelow, who made his fortune in real estate, founded Bigelow Aerospace in 1999 and promptly licensed the NASA patents for TransHab with the goal of developing orbital modules for sovereign customers.

The TransHab technology was originally pursued by NASA as a possible addition to the ISS but, due to funding shortfalls, the project was cancelled in 2000. Since then, Bigelow has developed the technology for his own purposes which include orbiting a space station in the not-too-distant future. The company is principally interested in sovereign customers and companies who will lease the habitats for research and science. One of the first steps towards realizing that goal was the launch of the unmanned prototype Genesis I and Genesis II modules that were flown in 2006 and 2007. They're still up there and doing very well. The next step is the launch of a BEAM on board a Falcon 9 to the ISS in 2016. The BEAM (Figure 8.11) is a scaled-down version of the BA-330 that Bigelow envisions orbiting as soon as SpaceX or Boeing man-rate their vehicles.

The cost? A flight on board the Dragon V2 to a BEAM will cost US$26.25 million – a figure that includes up to 10 days on board the BA-330. For those who want exclusive use of the habitat, lease blocks for one-third of a module costs about US$25 million for

8.10   Mr. Bigelow. Credit: NASA

8.11   BEAM attached to the International Space Station (ISS). Credit: NASA

60 days. So, if one of Bigelow's sovereign[2] customers decided to send a sole astronaut for a mission utilizing one-third of a BA-330, that mission would cost US$51.25 million. Of course, Bigelow's business plan depends on a man-rated vehicle being available to ferry his customers, which is why he takes a keen interest in the development of Dragon V2. And, despite the talk of Bigelow's habitats becoming space hotels (Bigelow owns Budget Suites of America), Bigelow has insisted that his company is not in the space-tourism business. Having said that, the ISS has become a popular tourist destination for those with US$40 or US$50 million lying around, so there may be some who eye the capacious BEAMs as an alternative.

DRAGON V2 DEVELOPMENT

But, before Dragon V2 can start ferrying astronauts and sovereign customers to orbit, SpaceX has a number of certification milestones it has to meet as part of the Commercial Crew Program. The first component of this is a series of reviews, of which the certification baseline review is the first. This review defines the steps SpaceX must take to meet NASA's safety requirements. Once this is checked off, SpaceX can submit Dragon V2 for the ISS design certification review, which assesses the vehicle for crew safety. After the ISS design certification review is the flight-test readiness review that demonstrates Dragon V2 is ready to execute its demonstration mission to the ISS while carrying one NASA astronaut. Assuming the flight to the ISS goes without a hitch, Dragon V2 is subject to the operational readiness review, which is almost the final check before approving that Dragon V2 is ready for operational status. Finally, there is the certification review in which NASA checks the vehicle one last time and clears the spacecraft for flights carrying astronauts and cargo. That's the first component. The second component, which was mentioned earlier in this chapter, are the missions. Remember, Dragon V2 must fly between two and six missions, carrying four astronauts on each flight. And there is a third component that NASA categorizes as "special studies" but there are few details about this, although it is possible this component could require specific orbital demonstrations beyond those required to dock and undock from the ISS.

Is there enough money to support this? Well, NASA's Fiscal Year 2015 budget request listed US$3.4 billion for commercial crew funding until 2019, which means there is a shortfall given that SpaceX and Boeing were awarded a total of US$6.8 billion. Where will that extra money come from? Perhaps there won't be any more funding within that time frame, in which case NASA may extend the contracts beyond 2019. Another option may be to use another internal program such as ISS operations perhaps? Or perhaps NASA will ask Congress for more money. Yet another option is that Dragon V2 and the CST-100 don't fly as many missions as intended, which would mean they would earn less of the funding pie. Or perhaps a composite of these options may be implemented.

---

[2] Bigelow has already signed memoranda of understanding with seven sovereign customers: Australia, Dubai, Japan, The Netherlands, Sweden, the United Arab Emirates, and the UK.

What we do know is that, for the five years leading up to 2015, Congress has provided US$1 billion fewer dollars for commercial crew programs than requested by NASA. That isn't a good trend.

> "SpaceX designed the Dragon spacecraft with the ultimate goal of transporting people to space. Successful completion of the Certification Baseline Review represents a critical step in that effort – we applaud our team's hard work to date and look forward to helping NASA return the transport of U.S. astronauts to American soil."
>
> *Gwynne Shotwell, SpaceX President and Chief Operating Officer*

In the meantime, SpaceX and Boeing are knuckling down to meeting those milestones. By late 2015, SpaceX had already checked off the certification baseline review and had completed the first of two pad abort tests (one from Cape Canaveral, Pad 40, and one from Vandenberg). Assuming the second pad abort test goes as planned, SpaceX will then set their sights on flying an unmanned Dragon V2 to the station in 2016 followed by a crewed flight sometime in 2017. Once that happens, NASA can finally stop having to stress about flying crews on the Soyuz and refocus on maximizing time on orbit, which means utilizing the facilities on board the ISS..

> "I hope I never have to write another check to Roscosmos."
>
> *NASA administrator Charlie Bolden during the media event*

With SpaceX's plan to make the Dragon V2 the most reliable manned spacecraft ever, the company may soon realize the goal of NASA's Administrator but, for Musk, the real goal is much further away – about 56 million kilometers away to be exact.

9

Red Dragons, Ice Dragons, and the Mars Colonial Transporter

Artist's concept of the Mars Colonial Transporter (MCT). Credit: Lazarus Luan

Musk wants to go to Mars, but not just for a flags-and-footprints mission. He wants to establish a colony of up to 80,000 – that's right: *eighty thousand* – people for the cost of about US$500,000 a ticket. To begin such an ambitious enterprise, a group of pioneers will travel to Mars boosted by a very, *very* big rocket fueled by liquid oxygen and methane. These Martian colonists – up to 100 of them – will travel in a very big spaceship called, appropriately enough, the Mars Colonial Transporter (MCT). But the MCT likely won't be the first SpaceX vehicle to land on the Red Planet. The first of these vehicles is likely to be an unmanned mission that will make use of a Dragon variant – Red Dragon. And,

© Springer International Publishing Switzerland 2016
E. Seedhouse, *SpaceX's Dragon: America's Next Generation Spacecraft*,
Springer Praxis Books, DOI 10.1007/978-3-319-21515-0_9

following Red Dragon, there is just the slightest possibility that a certain one-way mission may make it to the surface using off-the-shelf Dragons. We'll discuss these projects first before turning our attention to the MCT and the prospect of invading Mars.

RED DRAGON

Red Dragon is a variant of the Dragon vehicle that NASA is considering sending to the Red Planet as part of a sample return mission slated for 2022. The mission, which would also serve as a precursor to an eventual manned mission to Mars, has been on the drawing board of NASA's Ames Research Center (ARC) for a while. One of ARC's goals has been to drill into the Martian subsurface to search for life. In 2010, a researcher at Ames reckoned a modified version of Dragon V2 could probably do the job of landing on Mars and performing the drilling mission. After a few years studying the Red Dragon (Figure 9.1) concept, the Ames scientists agreed such a vehicle could execute a propulsive entry, descent, and landing (EDL), land two tonnes of useful payload, and possibly enable a sample return mission.

One of the benefits of using a repurposed Dragon is cost because all that is required are a few modifications. Also, the Dragon V2 is being developed to support propulsive landings, which is a key element of a Mars sample return mission. If SpaceX can get those SuperDraco engines working as they envisage, then Red Dragon will be able to land on Mars without the requirement for parachutes. SpaceX is also developing the Falcon Heavy, which may be the rocket that launches a Red Dragon on its journey. The Ames scientists reckon a modified Dragon V2 can deliver at least one tonne of payload to the surface at most sites on the northern plains and Hellas (Figure 9.2). Assuming such a mission is given the green light, one of the first tasks would be to get rid of the systems that are not required for an unmanned mission. This would mean jettisoning the crew

9.1 Red Dragon. Credit: NASA

9.2 Hellas region on Mars. Credit: NASA

systems and docking hardware, and upgrading the vehicle with deep-space communica-tion equipment, planetary protection systems, and payload access to the surface. Red Dragon would launch on a Falcon Heavy (Figure 9.3) and head for Mars. Any trajectory correction maneuvers en route could be dealt with by the vehicle's existing propulsion system. Thanks to its robust heat shield, Red Dragon wouldn't have a problem decelerat-ing through the Martian atmosphere and would land propulsively. Of course, there is the question of the vehicle's ballistic coefficient as defined by the formula $\beta = M/CdA$ (M is the mass of the Red Dragon, Cd is the vehicle's drag coefficient, and A is the aerody-namic reference area and the spacecraft's lift-to-drag ratio). A high ballistic coefficient means a high entry speed and aerodynamics can only do so much to slow a vehicle down. At some point, another system has to be used to bleed off the remaining speed. In Mars missions to date, those systems have included parachutes, but this won't be an option for Red Dragon because its cousin – Dragon – has been designed for Earth re-entry, which means it has a higher ballistic coefficient than Mars vehicles to date. In fact, Dragon's ballistic coefficient is right at the limit of what is possible when using parachutes, so deceleration and landing will have to be achieved via retro propulsion. Of course, this will require some tweaking of Red Dragon's propulsive capability and possibly fine-tuning the payload limits but, if the Ames scientists can make it work, it will be a major step towards solving the EDL problem for manned missions.

If the mission goes ahead, Red Dragon would probably be launched in 2022 and the vehicle would carry the equipment needed to return samples to Earth. This would require a Mars Ascent Vehicle and an Earth Return Vehicle (ERV). In NASA's mission plan, the Red Dragon ERV would be loaded with samples collected during an earlier rover mission that is tentatively planned for 2020.

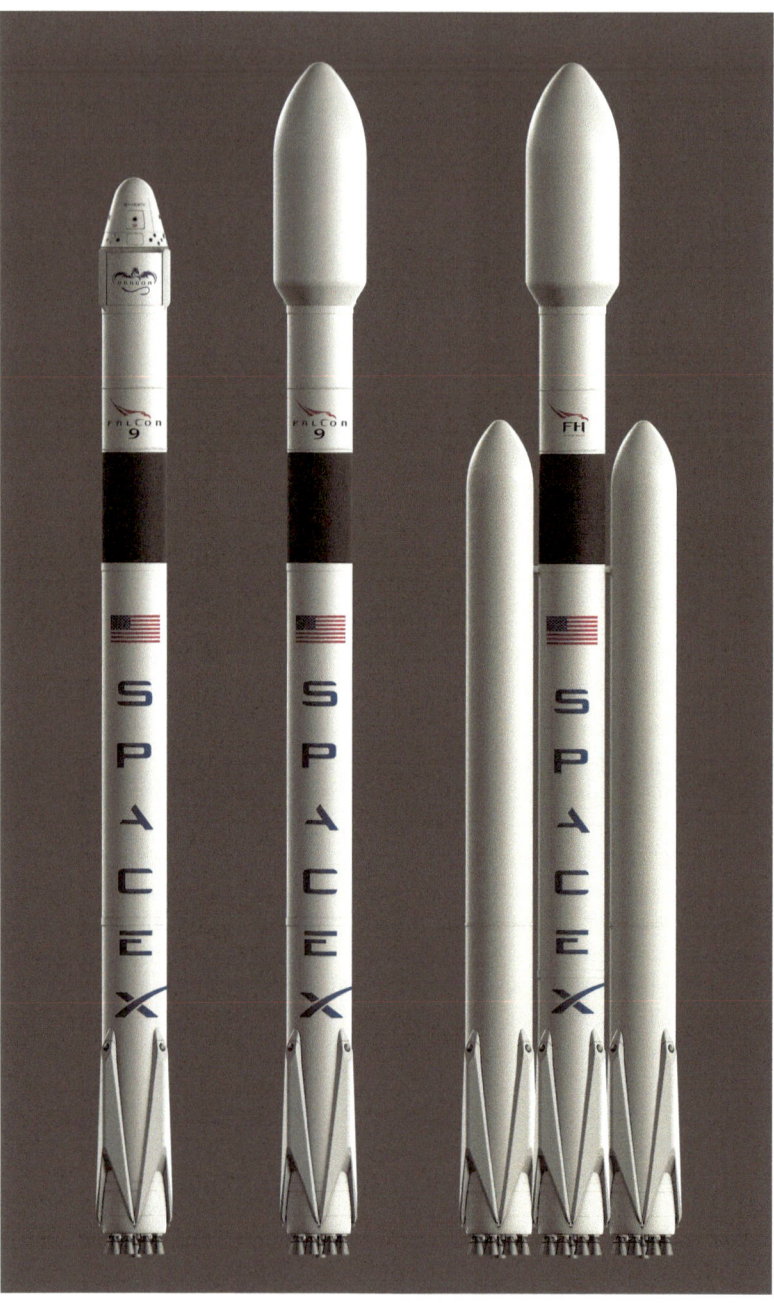

9.3   Falcon Heavy (R). Credit: SpaceX

ICE DRAGON

Another popular Mars mission is Ice Dragon, which uses a repurposed Dragon to search for life on the Martian surface and also to assess dangers to human missions. Ice Dragon fits neatly into NASA's Science Mission Directorate Astrobiology Program that has listed the search for life as its highest priority. Imagine if a Dragon was to find evidence of life on Mars: that would almost surely kick-start a manned mission. And, if a manned mission is green-flagged, then it would be helpful to know what resources – such as ground ice – are available to sustain a human presence. Ice Dragon (Figure 9.4) aims to answer those questions by using a modified Red Dragon.

In addition to being a helpful resource for humans (water and raw material for rocket fuel), ice is also important for other reasons. First, ice on Mars is found in the subsurface which is protected from the radiation environment and therefore any organic compounds should be fairly well preserved. Any ice will also provide a source of water to sustain biological activity. Since subsurface ice has been detected at the Phoenix landing site and at Amazonis Plantitia, it is possible the Ice Dragon vehicle will land at one of these locations [1]. Once there, it will have the following objectives:

1. Determine if life ever existed on Mars. Ice Dragon will do this by executing two strategies:

   (i) Search for biomolecules that provide evidence of biological activity.
   (ii) Search for simple organic molecules that may be related to biological processes.

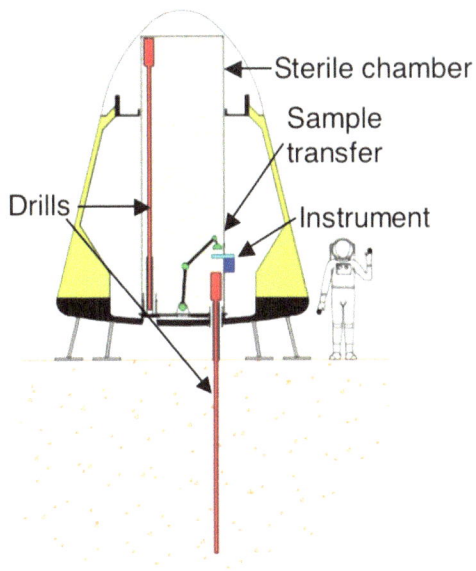

9.4 Ice Dragon concept. Credit: SpaceX

2. Assess subsurface habitability. Once on the surface, Ice Dragon will deploy a rotary percussive drill which will begin cutting down to a depth of two meters into the ice. As samples are cut, they will be placed into sample-acceptance ports.
3. Establish the origin and composition of ground ice. This will be achieved by imaging ice inside a borehole.
4. Define human hazards such as dust and cosmic radiation.
5. Demonstrate in situ resource utilization (ISRU).
6. Execute human EDL demonstration.

RED DRAGON TO MARS

To launch a Red Dragon or an Ice Dragon to Mars will require a big launch vehicle, but SpaceX is working on that by developing the Falcon Heavy (Table 9.1). Powered by 27 Merlin engines arranged in three Falcon 9 cores, the Falcon Heavy features reusable technology, since the two side boosters and core stage will propulsively land [2]. The impetus for creating the Falcon Heavy was to compete with the world's heavy launchers such as the Delta IV Heavy and the Ariane 5. The first performance data for the Falcon Heavy were revealed in 2006. These showed the Falcon Heavy as having a low Earth orbit (LEO) capability of 24,750 kilograms at a cost of US$78 million. As the launcher developed, these numbers changed, as did the date for the first launch, which was originally planned for 2013. Using 2015 data, the Falcon Heavy can ferry 53,000 kilograms into LEO, 21,200 kilograms to geostationary transfer orbit (GTO), and 13,200 kilograms can be placed in a Trans-Mars Trajectory.

Standing nearly 70 meters tall with a launch mass of nearly 1,500,00 kilograms, the Falcon Heavy comprises three cores, with propellant being cross-fed between cores for payloads weighing more than 45 tonnes. For space aficionados, the central core stage is almost the same as the Falcon 9 v1.1 (Appendix VI). Standing 43 meters tall and measuring 3.66 meters in diameter, this stage can carry approximately 414,000 kilograms of propellant and, like the side boosters, is powered by the workhorses of the SpaceX launch enterprise: the open-cycle Merlin 1D. The Merlin 1D generates a sea-level thrust of 66,700 kilograms and

Table 9.1 Falcon Heavy by the numbers

Height	68.4 m
Core diameter	3.66 m
Launch mass	1,462,836 kg
Stages	2
Boosters	2
Span	11.6 m
Mass to low Earth orbit (LEO)	53,000 kg (28.5°)
Mass to geostationary transfer orbit (GTO)	21,200 kg (27°)
Payload to Mars	13,200 kg
Total lift-off thrust	17,615 kN
Total Vac thrust	20,017 kN

a thrust in a vacuum of 73,000 kilograms. Designed with a deep throttling capability (70% to 112% of rated performance), the Merlin 1D enables a flexible mission profile and, thanks to the on-board re-ignition systems on the first stage, that aforementioned propulsive return capability is also available. In common with its older cousin, the Falcon 9, the Falcon Heavy sports an engine-out capability through most of the first-stage flight.

Now to the boosters. In an effort to reduce system complexity, SpaceX has designed its Falcon Heavy launcher with as much commonality as possible. This means that the designs of the core stage and the boosters are very much alike, which in turn means that the manufacturing process can use the same tools and techniques. While commonality saves money and time, it is nothing new in the launch industry (the principle is applied to the Delta IV Heavy and the Angara), but the notion of reusability is, and this is one of SpaceX's biggest goals: following a sequence of propulsive maneuvers, the three cores of the Falcon Heavy will return to the ground for a soft landing either near the launch site or on the Autonomous Spaceport Drone Ship (ASDS) (Figure 9.5), which has been used to test returning Falcon boosters.

When operational, the Falcon Heavy would launch and burn the two boosters for almost three minutes, after which the boosters would detach from the core and head for home. To execute this return maneuver, the boosters would first swivel to an engines-first orientation in preparation for a boost-back burn. This burn would reduce the downrange speed and enable the boosters to make their way back to the launch site or the spaceport. At 70 kilometers altitude, the boosters would fire up three engines to slow the booster and, as the boosters began their descent, they would use four grid fins to steer and stabilize the booster during the remainder of the descent. The grid fins, which can perform at supersonic and subsonic speeds, can rotate and tilt simultaneously, enabling very precise guidance and control during flight. And this precision will be needed if the boosters are to land with the pinpoint accuracy required to land on a platform. As the boosters made their flight corrections, their flight path would be modified to ensure the boosters were on target for the landing site. At 28 seconds before landing, the booster's center engine would re-ignite for the landing burn and, 10 seconds before landing, the legs would deploy. Landing would be at about six meters per second.

In addition to the core and the two boosters, the Falcon Heavy uses a Falcon 9 v1.1 second stage (Appendix VI) which will be about 14 meters long and weigh a little less than five tonnes. The diameter will be the same as the core stage and its fuel load will be 97,000 kilograms. Powered by a single Merlin 1D Vac engine, the second stage is fitted with a reaction control system that enables three-axis control during the coast phase.

MARS ONE AND RED SUPERDRAGONS

"'It looks like a scam. They don't have any technology, they don't have any agreements with the space industry. It looks very shaky.' The bigger problem? Mars One's flaws, too few spaceships, nonexistent life-support technologies, not nearly enough money, and, really, no good reason for going discredit all Mars exploration plans, including NASA's."

John Logsdon, space policy expert at George Washington University
in Washington, DC

9.5 "X" marks the sport! An overhead shot of SpaceX's Autonomous Spaceport Drone Ship (ASDS). Returning a spacecraft or any vehicle from space is a major challenge, but returning that vehicle to a precise landing is even more of a headache, especially when the vehicle – the Falcon 9 first stage – in question is traveling up to 1,300 m/sec. To achieve a precision landing, the first goal is to stabilize the stage. Once that's done, the next step is to reduce speed. This is achieved by relighting engines in a series of burns. The boost-back burn is the first. This aligns the impact point of the stage. The boost-back burn is followed by the supersonic retro propulsion burn that helps to bleed off that speed to around 250 m/sec. Finally, the landing burn is executed, which bleeds off a lot more speed so the stage is moving at only 2 m/sec once the legs deploy for landing. Sounds fairly simple, but it is anything but because the landing site measures less than 30 by 10 m. Plus that landing site isn't perfectly stationary. It is a ship after all! But, if SpaceX can perfect this style of landing, it will revolutionize how rockets land on Earth and significantly reduce the cost of getting into space. Credit: NASA

Why is there a section on Mars One in a book about SpaceX's Dragon? Well, Mars One's mission architecture, such as it is, seems to rely heavily on the vehicle judging by the promotional video and images on the organization's website (Figure 9.6). As you can see from the quote above, reactions to Mars One have been … colorful. Reality television junkies are excited whereas those in the space community are, for the most part, very, *very* skeptical.

For those who haven't been following the Mars One story, the Dutch not-for-profit venture is the creation of Bas Lansdorp, and the idea is that the establishment of the first colony on Mars will be broadcast as a reality television show. Here's the nuts and bolts of

9.6 Artist's concept of a Mars One lander making its way to the surface of the Red Planet. The mission is a long shot at best but, if it does go ahead, repurposed Dragons may be used to transport cargo and crew to Mars. Credit: Mars One

the mission. Beginning in 2018, Mars One will begin sending cargo missions to the Red Planet to deliver all the bits and pieces the reality TV stars will need when they arrive. Assuming all the cargo arrives safely, the Mars One astronauts will start blasting off from Earth sometime around 2024. Once there, the first Mars colonists will set up base with the knowledge that none of them will be returning to Earth. For many of the skeptics, this is a good thing. Every 26 months after the first human mission, another crew will leave Earth en route to Mars. Gradually, the Mars One base (Figure 9.7) will grow into a self-sustaining colony. All funded by reality TV revenue! Sane or insane? You decide, but let's begin with the technology, because some of the technology may well include a Dragon or three. Why do we know this? Well, for one thing, the Mars One website is littered with images of vehicles that look suspiciously like Dragons and, for another, Mars One claims that their plan is built on existing technologies (Dragon, Falcon) that are available from proven suppliers (SpaceX). Mars One won't be building any of the hardware: it will buy it from third parties. Let's examine that claim. In 2015, there are no vehicles capable of ferrying humans to the Red Planet, although SpaceX is working hard to change that, as we will see shortly. Also, we know we don't have the technology to safely land humans on the surface of the Red Planet. Again, SpaceX is pioneering the solution to that problem. That's the technical side of the venture. The human side? Well, if you happen to be photogenic and have a sense of humor suited for reality television, then you have a good chance of being selected, because, as we know from decades of experience of selecting astronauts, good looks and the ability to laugh are critical characteristics when it comes to surviving austere environments! The cost of this ambitious mission is estimated to be around the US$6 billion mark. How Mars One will get this funding is just one of many unknowns of the mission, but let's return our focus on the Dragon-related technology.

9.7 If – and it's a big, big "if" – Mars One is able to get the US$6 billion they need for their mission, this is what their base may look like: a string of Dragons. Credit: Mars One

We'll begin with the transit vehicle, which may be a variant of Dragon launched atop a SpaceX Falcon Heavy. As you can see in Figure 9.7, a number of these vehicles will be linked together to form the Mars One base. Some of the Dragons will serve as life-support units, others will be dedicated as supply units, and others will serve as living units that may be attached to inflatable habitats. In the computer-generated images, Mars One does a good job convincing the man in the street (and the Mars One applicants) that this colony can be realized, but there are more than a few flaws, one of which was highlighted in a 2014 MIT study that calculated the first colonist would die after 10 weeks due to excess oxygen levels.

> "It will take quite a bit longer and be quite a bit more expensive. When they first asked me to be involved I told them 'you have to put a zero after everything'."
>
> *Gerard 't Hooft, a Dutch Nobel laureate and ambassador for Mars One,*
> *who said he did not believe the mission could take off by 2024 as planned*

Gerard 't Hooft may be correct (Nobel laureates usually are). After all, establishing the first colony on the Red Planet Mars One style will require as many as 15 Falcon Heavy launches with a bill of about US$4.5 billion. Compare that number with the almost US$16 billion it will cost NASA to fly just one manned Orion flight, and we're not even considering the cost of the Space Launch System. Financially implausible, risky architecture, and some downright dodgy life-support assumptions all conspire to make the Mars One venture seem a little far-fetched, so let's turn our attention to the company that probably has the best plan to get us to the Red Planet.

MARS ON A BUDGET

When you read about Mars missions that cost several billions of dollars, you may be wondering how Musk can sell a ticket for US$500,000, which is just a tenth of the cost of Sarah Brightman's 2015 trip to the International Space Station (ISS). But Musk reckons his colony program will be funded by governments and private enterprise, pointing to historical precedents of the British establishing colonies in North America. How will he do it? Well, we're not sure, but a major element is the development of that very big rocket that will be able to take off and land vertically. Other key factors will be the development of a powerful combustion engine and a crew vehicle. A Dragon? Well, we've discussed Red Dragon in this chapter, but Musk has another vehicle in mind – the MCT. Another factor in Musk's favor is that he is not a government enterprise. Consider for a moment NASA's Design Reference Mission (DRM): the Mars Design Reference Architecture 5.0.

NASA's Mars plans

NASA has been designing Mars missions for a while. Its first DRM, designated DRM 1.0, was developed in 1992–93 following President George H.W. Bush's short-lived Space Exploration Initiative (SEI), which was kick-started in 1989. DRM 1.0 was based on Martin Marietta's Mars Direct mission, but work on the plan ground to a halt at once following the termination of SEI. The next DRM was created in 1996 following the discovery of ALH 84001, a Martian meteorite that was thought to contain microfossils. This DRM was designated DRM 3.0 (no DRM 2.0 exists), but it was also known by another acronym derived from the intended mode of propulsion. DRM 3.0 was a nuclear mission based on Bimodal Nuclear Thermal Rocket (BTNR) propulsion (Figure 9.8). DRM 3.0 was a true *Battlestar Gallactica* approach to transporting humans to Mars and never gained much momentum, although it did spawn DRM 4.0, which was the BTNR architecture with the added feature of a dual lander. DRM 4.0 was then modified to be in line with the plans for the Constellation Program and, when Constellation died, DRM 4.0 metamorphosed into DRM 5.0. Let's take a look.

DRM 5.0

In common with previous DRMs, DRM 5.0 was compiled by a very select and esteemed panel. The 83-page document makes for interesting and difficult reading, and also provides a cautionary (and downright frightening) insight into the way government agencies work. The mission was designed to use nuclear thermal rocket (NTR) propulsion, although there was another version that used chemical propulsion. This propulsion mode was a strange choice because there are no funding dollars available to fund a NTR program and, with the way things are going with the Space Launch System (SLS), there won't be any funding for quite a while. Then there was the messy process of getting all the bits and pieces into LEO. In the original DRM 5.0, the mission elements would have been launched on the now defunct Ares V, which has now morphed into the SLS. If NASA were to pursue

9.8 Getting to Mars safely is all about getting there fast. Very fast. And one way to do this is to use a propulsion system that really packs a punch. The Bimodal Nuclear Thermal Rocket (BTNR) generates huge specific impulse numbers (between 860 and 975 sec), which would mean the six-month trip using traditional chemical propulsion could be halved. Credit: NASA

the DRM 5.0 today, it would require about a dozen SLS launches to transport the 20 major vehicle elements into LEO. That's an awful lot of launches, and remember this is a government launch system, so such a marathon sequence of launches wouldn't come cheap. Once in orbit, each Mars mission would begin with two cargo vehicles that would be sent from LEO to a Mars transfer orbit. One of these cargo flights would include a Mars Ascent Vehicle (MAV), also known as the Descent-Ascent Vehicle (DAV), which would be used by the astronauts to return to Mars orbit. The second cargo flight would combine a surface hab and a lander for the crew. The cargo would be located in giant aeroshells which would serve a dual purpose as an aerocapture system into Mars orbit. As for the crew vehicle, it is unlikely this could ever perform an aerocapture because it is too big, which would mean a very expensive propulsive capture would be required. DRM 5.0 calls for three missions spread out over 10 years, with each mission landing in a different location. Reading through the DRM 5.0 document, it is impossible not to be struck by the marked lack of redundancy and the horrendous waste of components – just about everything is expendable.

On arrival in Mars orbit, one cargo remains on station while the other with the MAV descends to the surface propulsively. Surface power is provided by a mini nuclear reactor which is off-loaded from the MAV robotically. Twenty-six months after the cargo launches, the crew departs LEO in the Mars Transfer Vehicle (MTV). Comprising an Orion, three Trans-Mars Injection (TMI) stages, a Mars Orbit Insertion (MOI) stage, and a Trans-Earth Injection (TEI) stage (all expendable by the way because this was a "money no object" mission apparently), the spacecraft is launched on its way to Mars carrying six crew. On arriving at Mars, the MOI is used to propulsively enter Mars orbit. No aerocapture required

9.9 Mars the NASA DRM 5.0 way. Credit: NASA

evidently. The astronauts rendezvous with the lander/hab module and descend to the surface (Figure 9.9) in the same way as the first cargo payload did.

As you can see in Figure 9.9, the crew's living quarters are perched on landing legs, so we're forced to assume this crew is particularly radiation-resistant, especially since they would have a 500-day stay ahead of them. While the three missions at three locations is a good idea for exploration, it adds to the cost because there are no backups or duplicates of essential equipment. At the end of the mission, assuming they haven't succumbed to radiation sickness, the crew clambers on board the MAV, returns to Mars orbit, and docks with the MTV in preparation for the trip to Earth. Finally, after 900 odd days, the crew splashes down in the Orion, which is the only element that survives the mission. To fly out, the three missions would require 36 SLS launches to LEO to orbit 60 mission components. All expendable remember! Deficiencies? Yes, there are a few. For one thing, the mission design is not progressive and, for another, there is no explanation of how all the bits and pieces will be assembled in LEO. And then there is the cost of all that fuel. Several thousand tonnes of fuel. Then there is the small matter of redundancy. If you happen to be an astronaut landing on a planet millions of kilometers from home, then one level of redundancy that is nice to have is engine-out capability, but the DRM 5.0 crew won't have that. Another redundancy feature is designing the crew vehicle so it can be used as an escape vehicle in case things go pear-shaped. For those old enough to remember the Apollo era, you will remember that this capability was present in the lunar module during descent. Bear in mind that, since parachutes don't work very well on Mars, the crew is completely

Table 9.2 The incredible expendable Mars mission.

Mission element	Mass (tonnes)
Total mass placed in low Earth orbit (LEO)	1,251.8
Total mass of orbit re-boost modules (for use in LEO only)	−106.4
Total expedition mass in LEO	1,106.4
Total mass of propulsion vehicles and fuel used for Trans-Mars Injection (TMI)	−779.0
Total post-TMI vehicle and fuel mass sent into Mars transit	366.4
Total post-TMI in-space propulsion vehicle and fuel mass	−108.5
Total expeditionary mass arriving in Mars orbit less propulsion	258.0
Mass departing orbit for Mars surface	206.0
Payload mass landing on Mars surface	80.8

reliant on rocket power if they are to survive any anomaly. Apparently, the notion of redundancy didn't get much of a look-in when DRM 5.0 was being designed. Another shortcoming is the miniscule mass (Table 9.2) that is landed on the surface. Just two vehicles land, remember, and much of the mass of vehicle is propellant. And then there's the design of the crew habitat which can't be buried to protect the crew from radiation.

If you look at Table 9.2, it is difficult not to be reminded of a similar mission architecture that started with a massive rocket and ended with a tiny capsule. This was how NASA flew to the Moon! But that was nearly five decades ago! Surely, with all the new technology we have nowadays, mission planners can come up with a more efficient and safe way to get to Mars. Can't they? There will probably be a DRM 6.0 at some point and we can only hope that those around the table learn from the shortcomings of DRM 5.0. If they do, then DRM 6.0 should be based on a reusable launcher and spacecraft. It should also provide redundancy to ensure crew survival and include engine-out capability during descent and ascent. It should also provide some way of protecting crews from radiation. And it should incorporate some vision and that vision has to be creating an architecture that expands the human footprint in space and this can only be achieved by developing reusable architectures. DRM 5.0 is a real eye-opener to the realities of a government-funded manned Mars mission, because that reality is that such a mission will never happen. Ever. Don't blame NASA. This is an agency that has been decimated by funding and personnel over the years. Since Apollo, NASA has suffered a nearly 50% reduction in budget as a percentage of federal spending. For those ardent supporters of a manned mission to Mars sooner rather than later, your very best hope is not a reality television show or a bloated government venture, but a private company that has been achieving great things in the commercial spaceflight arena of late. Its name is SpaceX.

COLONIZING MARS THE SPACEX WAY

While the mission architecture for SpaceX's colonization of the Red Planet has yet to be defined, we do know that Musk will need a very big rocket powered by a very, *very* big and powerful engine. We'll take a look at the engine first. As with all of SpaceX's rockets and spacecraft, this engine has a cool name – the Raptor.

Raptor power

One of the first pieces of SpaceX's Mars puzzle is the Raptor, destined to be one of the most powerful rocket engines ever created. While the Space Shuttle Main Engine (SSME) delivered 375,000 pounds of thrust, the Raptor will deliver one million pounds, a number exceeded only by the mighty Saturn V F-1 engine. The key to developing such phenomenal thrust numbers is SpaceX's decision to use a full-flow cycle design. This design, which sends fuel through turbopumps, has never been attempted in the US, although the US Air Force and NASA have considered the full-flow cycle option over the years. The advantage of channeling all the fuel through turbopumps is that the fuel and oxidizer passing through the preburners drive the turbopumps more forcefully. This in turn increases the pressure in the combustion chamber, which translates into higher engine performance.

SpaceX is developing the Raptor at NASA's Stennis Space Center. Assuming the development goes well, Musk reckons the engine could be ready in the mid-2020s time frame. Is he being realistic? Well, SpaceX has always been very innovative when it comes to designing rocket engines. They have also been extremely aggressive when it comes to development and cost. But, as advanced as their current crop of engines are, the technology, while innovative, has never been unproven, which is why the Raptor, when it was

Parameter	Value
Propellants	LOX/LH2
Vacuum Thrust	150,000 lbf
Vacuum Isp	470.1 seconds
Chamber Pressure	1700 psia
Nozzle Area Ratio	250:1
Mixture Ratio (Ox/fuel)	6.0
Throttle Range	50 - 100 %

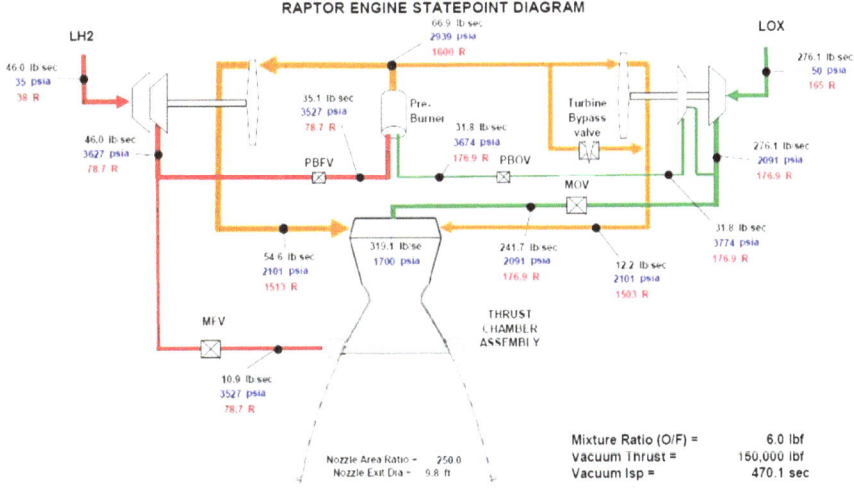

9.10   Raptor Engine Statepoint diagram. Credit: SpaceX/NASA/Markusic

presented as a concept in 2010, is a clear break from SpaceX's tradition of designing engines. Not only is developing the full-flow cycle engine stepping into the arena of unproven technology, it is also represents a departure from the low-cost ethos that has defined SpaceX for so long. But, if you want to get to Mars, you have to pay a price and that price will be the development of the know-how of producing a full-flow cycle engine. Fueled by liquid methane and liquid oxygen, the Raptor, when operational, will be the rocket that powers the MCT – of which more later – to Mars. Can SpaceX develop an unproven engine in such a short time frame? Perhaps. After all, the rocket engine is not completely unproven. Two full-flow engines have appeared on test stands: the RD-270 and AeroJet's Integrated Powerhead Demonstrator. The RD-270 was the engine that was developed to help the Soviets reach Mars, so in that sense the objective is similar to the development of the Raptor. The RD-270 was a single-nozzle engine that delivered 1.4 million pounds of thrust at sea level. Between October 1967 and July 1969, 22 prototypes performed 27 firings, only nine of which were nominal, which is why the program was cancelled. A slightly less powerful full-flow cycle engine was the Integrated Powerhead Demonstrator, which delivered 250 pounds of thrust. A joint DoD/NASA program, this engine was developed to test technologies that could result in an engine with a higher performance than the SSME and one that could be used for up to 200 missions.

> "The Mars transport system will be a completely new architecture. I am hoping to present that towards the end of this year. Good thing we didn't do it sooner, as we have learned a huge amount from Falcon and Dragon."
> *Elon Musk responding to a question during an "Ask Me Anything" chat session on the online-aggregation site Reddit, January 2015*

MARS COLONIAL TRANSPORTER

The MCT is one of the key elements in Elon Musk's plan to establish a colony on the Red Planet. Designed to carry as many as 100 astronauts to the surface of Mars, the MCT builds on the development, technology, and experience of Dragon. The details of the MCT mission architecture are sparse, but it is expected the first flights will carry fewer people because much of the volume will be needed to ferry supplies and equipment necessary to establish the outpost. The spacecraft is estimated to weigh 100 tonnes, which means the launcher of choice will be one of SpaceX's super-heavy lift launch vehicles.

In keeping with the SpaceX mantra of reusability, the MCT is designed to be reused, and it is also designed to serve not just as a passenger vehicle, but also as a means of getting cargo into orbit. One suggestion is to use the MCT to launch Bigelow's expandable modules into orbit where they could be linked together to create a colony ship capable of ferrying thousands of people. A company called Tethers Unlimited is currently working on the technologies required to build such a spacecraft on orbit. Funded by a US$500,000 NASA Institute for Advanced Concepts (NIAC) grant, Tethers Unlimited are working on developing trusses that can be used to link a number of Bigelow Expandable Activity Modules (BEAMs) to form one very big spacecraft. Such a large interplanetary spacecraft might just be the ticket to creating an outpost that could eventually become the self-sustaining colony that Musk has been working towards. Could it work? Well, SpaceX's track record in keeping costs low has been pretty good so far and that's one of the keys to the Mars colonization business. By amortizing the cost of the vehicle over many trips, the costs of this venture come down a little, but how could these costs be reduced to a minimum? One way will be choosing the right launch windows. We know conjunction launch windows are more suited to a manned mission because they enable a 500-day stay on the surface, whereas an opposition launch window permits only a month on the surface. For cargo purposes, the opposition launch window is ideal because you want to get as much cargo onto the surface as quickly as possible. After delivering the cargo, the MCT would return to Earth to collect its human payload, which it would transport to Mars on an opposition launch window, or it could collect another cargo payload and there could be a separate MCT dedicated for passengers. Returning to the subject of amortization, if there was just one cargo MCT, then that spacecraft could perform a round trip every 13 months, but perhaps the mission architecture could be tweaked a little by having the cargo MCT stay in space and have the cargo passed to another vehicle for the descent to the surface? The truth is we don't know what mission architecture Musk has in mind.

SPACEX'S BFR

Just as the Falcon Heavy is at the heart of any program to send Red Dragons to Mars, the development of SpaceX's BFR (Big Falcon Rocket – this is the name that can be used publicly) is key to sending the MCT to the Red Planet. When this behemoth is built, it will dwarf even the mighty Saturn V. Capable of launching 150 tonnes of payload, the BFR will out-launch even the SLS. SpaceX estimate the development cost of this colossal rocket will be around the US$2.5 billion mark, which is approximately 15 months of SLS funding. There may some reading this wondering why Congress is asking NASA to build the SLS (a less capable system) when the BFR can do the same job for less money. Well, NASA suggested cancelling the SLS, but Congressional pork-barrel politics wouldn't let them, so the "rocket to nowhere" (the system is so expensive that there will be no money left to build payloads to fly on the thing) was born and the American taxpayer was once again fleeced. But let's get back to the monster rocket. We know the BFR will be powered by nine Raptors but, beyond this, the plans are a little sketchy. We know the Raptor engines will be used on a 10-meter-diameter core and we know the vehicle can be scaled up to three cores, which would result in a BFR with 27 Raptor engines. We also know that SpaceX may increase the size of these cores to 12.5 meters and even 15 meters if the decision is made to utilize just a single core. And we know this BFR (it doesn't have an official name yet) will be capable of sending the 100-tonne MCT to Mars. What would such a vehicle look like? Let's indulge in some hypothetical thinking. First, we need a starting point, so let's use the Falcon 9 v1.1. The first thing we have to do is scale up the Falcon 9 to the thrust levels of the Raptor. We know the Falcon 9 v1.1 first-stage weight is 404 tonnes and we know the second-stage weight is 99 tonnes. To do the conversion, we multiply the weights by the ratio between the thrust of the Raptor (specific impulse of 321 seconds at sea level) and the thrust of the Merlin 1D which powers the Falcon 9 v1.1. This results in a gross lift-off weight of 2,452 tonnes for the first stage of our reference rocket and 582 tonnes for the second stage. Now another ratio calculation must be applied because we know that the stages will have a propellant mass fraction of around 0.94; 2,452 multiplied by 0.94 leaves us with 2,305 tonnes and an empty mass of 147 tonnes for the first stage. Add the first and second stages together and you have a rocket that weighs 3,044 tonnes without the payload. That's a BFR that will launch 145 tonnes to LEO and 22 tonnes to Mars. But, if SpaceX wants to make its BFR reusable, some of the payload to LEO has to be trimmed because some fuel will be needed to return the boosters to the launch pad. SpaceX have indicated that there is about a 30% reduction in payload per stage to achieve reusability which means the new reusable BFR will be capable of launching around 110 tonnes to LEO. But SpaceX needs to get that 100-tonne MCT to Mars so an even bigger rocket is needed which means a tri-core BFR will most likely be required. Applying the same principle of reusability, such a rocket could launch around 230 tonnes to LEO and just under 40 tonnes to Mars – impressive, but still not enough. Another way to increase payload would be to reduce the propellant load and have the boosters parachute back rather than use propulsion. That would increase the payload to Mars to about 60 tonnes. And you could scale up the boosters and add a third stage, and that would probably be enough to get you to the magic 100-tonne number, although you would need to use two engines on that upper stage. The problem now becomes where could SpaceX launch such a monster rocket?

REFERENCES

1. Stoker, C.R.; Davila, A.; Davis, S.; Glass, B.; Gonzales, A.; Heldmann, J.; Karcz, J.; Lemke, L.; Sanders, G. Ice Dragon: A Mission to Address Science and Human Exploration Objectives on Mars. *Concepts and Approaches for Mars Exploration* (2012).
2. Karcz, J.S.; Davis, S.M.; Aftosmis, M.J.; Allen, G.A.; Bakhtian, N.M.; Dyakonov, A.A.; Edquist, K.T.; Glass, B.J.; Gonzales, A.A.; Heldmann, J.L.; Lemke, L.G.; Marinova, M.M.; Mckay, C.P.; Stoker, C.R.; Wooster, P.D.; Zarchi, K.A. Red Dragon: Low-Cost Access to the Surface of Mars Using Commercial Capabilities. *Concepts and Approaches for Mars Exploration* (2012).

Appendix I: Dragon C2/3 Cargo Manifest

SPACE STATION CARGO

1. Food and crew provisions: 306 kg
   - 13 bags standard rations: food, about 117 standard meals, and 45 low-sodium meals
   - 5 bags low-sodium rations
   - Crew clothing
   - Pantry items (batteries, etc.)
   - SODF and Official Flight Kit

2. Utilization payloads: 21 kg
   - NanoRacks Module 9 for US National Laboratory: NanoRacks-CubeLabs Module-9 uses a two-cube unit box for student competition investigations using 15 liquid mixing tube assemblies that function similarly to commercial glow sticks. Science goals for NanoRacks-CubeLabs Module-9 range from microbial growth to water purification in microgravity
   - Ice bricks: for cooling and transfer of experiment samples

3. Cargo bags: 123 kg
   - Cargo bags: reposition of cargo bags for future flights

4. Computers and supplies: 10 kg
   - Laptop, batteries, power supply cables

   Total cargo upmass: 460 kg (520 kg including packaging)

RETURN CARGO

1. Crew preference items: 143 kg
   - Crew preference items, official flight kit items

© Springer International Publishing Switzerland 2016

E. Seedhouse, *SpaceX's Dragon: America's Next Generation Spacecraft*,
Springer Praxis Books, DOI 10.1007/978-3-319-21515-0

2. Utilization payloads: 93 kg
   - "Plant Signaling" hardware (16 Experiment Unique Equipment Assemblies): Plant Signaling seeks to understand the molecular mechanisms plants use to sense and respond to changes in their environment. Ambient Hardware return only; no plant sample return (24 kg)
   - Shear History Extensional Rheology Experiment (SHERE) hardware: SHERE seeks to understand how liquid polymers behave in microgravity by measuring response to straining and stressing. Ambient hardware return; no samples (36 kg)
   - Materials Science Research Rack (MSRR) Sample Cartridge Assemblies (Qty 3): MSRR experiments examined various aspects of alloy materials processing in microgravity; SETA (Solidification along a Eutectic path in Ternary Alloys-2); MICAST/CETSOL (Microstructure Formation in Casting of Technical Alloys under Diffusive and Magnetically Controlled Convective Conditions/Columnar-to-Equiaxed Transition in Solidification Processing); Ambient hardware return with samples (9 kg)
   - Other: Supporting research hardware such as Combustion Integrated Rack (CIR) and Active Rack Isolation (ARIS) components, double cold bags, MSG tapes

3. Systems hardware: 345 kg
   - Multifiltration bed
   - Fluids control and pump assembly
   - Iodine compatible water containers
   - Japan Aerospace Exploration Agency (JAXA) multiplexer

4. Extravehicular activity (EVA) hardware: 39 kg

   Total cargo downmass: 620 kg (660 including packaging)

Appendix II: DragonLab™

© Springer International Publishing Switzerland 2016

E. Seedhouse, *SpaceX's Dragon: America's Next Generation Spacecraft*,
Springer Praxis Books, DOI 10.1007/978-3-319-21515-0

SPACEX

DragonLab™
Fast track to flight.

OVERALL DRAGON™ CAPABILITIES

Dragon is a free-flying, reusable spacecraft capable of hosting pressurized and unpressurized payloads. Subsystems include propulsion, power, thermal control, environmental control, avionics, communications, thermal protection, flight software, guidance, navigation & control, entry, descent & landing, and recovery.

USES

- Highly Responsive payload hosting
- Sensors/apertures up to 3.5m diameter
- Instruments and sensor testing
- Spacecraft deployment
- Space physics and relativity experiments
- Radiation effects research
- Microgravity research
- Life science and biotech studies
- Earth sciences and observations
- Materials and space environments research
- Rendezvous and inspection
- Robotic servicing

DRAGON SPACECRAFT SYSTEM

- Fully recoverable capsule
- Trunk jettisoned prior to reentry
- 6000 kg total combined up-mass capability
- Up to 3000 kg down mass
- Payload Volume:
  - 10 m³ pressurized
  - 14 m³ unpressurized
- Mission Duration: 1 week to 2 years
- Payload Integration timeline:
  - Nominal: L-14 days
  - Late-load: T-9 hours
- Payload Return:
  - Nominal: End-of-Mission + 14 days
  - Early Access: End-of-Mission + 6 hours

TYPICAL INTEGRATION TIMELINE

ATP   ICD   Fit Check   Payload Integration   Return

L: – 5 Mth   – 3 Mth   L-2 Mths – 2 Wks   EOM 2 wk

Launch

v.2.1

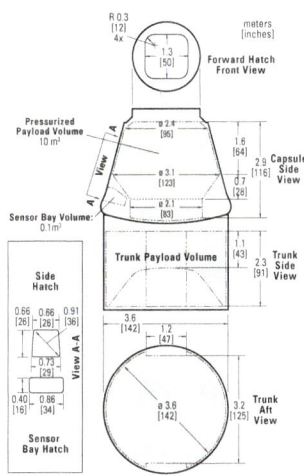

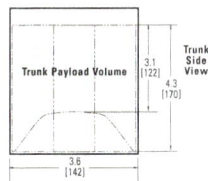

OPTIONAL TRUNK EXTENSION

SPACEX

For more information, please email us at DragonLab@spacex.com.

spacex.com

PAYLOAD SERVICES

MECHANICAL

- Specific mounting locations and environments are mission-unique
- **Pressure Vessel Interior** (pressurized, recoverable)
  - 10 m³ payload volume
  - Lab temp, pressure and RH
  - Typically Middeck Locker accommodations
  - Other mounting arrangements available
- **Sensor Bay** (unpressurized, recoverable)
  - Approx 0.1 m³ (4cu ft) volume
  - Hatch opens after orbit insertion; closes prior to reentry
  - Electrical pass-throughs into pressure vessel
- **Trunk** (unpressurized, non-recoverable)
  - 14 m³ payload volume
  - Optional trunk extension for a total of up to 4.3 m length, payload volume 34 m³

POWER

- 28 VDC & 120 VDC
- Up to 1500-2000 W average; up to 4000 W peak

THERMAL & ENVIRONMENTAL

(ref. NASA SSP 57000)

- Internal Temp: 10~46 °C
- Internal Humidity: 25~75% RH
- Internal Pressure: 13.9~14.9 psia
- Cleanliness: Visibly Clean–Sensitive (SN–C–0005)
- Pressurized: convective or cold-plate
- Unpressurized: cold-plates if required
- Payload random vibration environment:
  - Pressurized: 2.4 grms (> 100 lbm)
  - Unpressurized: 2.9 grms

TELEMETRY & COMMAND

- Payload RS-422 serial I/O, 1553, and Ethernet interfaces (all locations)
- IP addressable payload standard service
- Command uplink: 300 kbps
- Telemetry/data downlink: 300 Mbps (higher rates available)

SPACECRAFT SUBSYSTEMS

STRUCTURES AND MECHANISMS

- All Structures and Mechanisms are designed to be capable of supporting crew transportation, consistent with all relevant NASA standards and Factors of Safety
- 3 or 4 windows, 30 cm diameter
- Sensor Bay Hatch: deployable/retractable hatch mechanism which opens on orbit and closes prior to reentry
- Capsule/Trunk fluid & electrical interconnects

PROPULSION

- 12-18 Draco thrusters
- NTO/MMH hypergolic propellants
- Up to 2 fault tolerant

AVIONICS

- Dual/Quad fault tolerant Flight Computers
- Multiple generic Remote Input/Output (RIO) modules with customized complements of Personality Modules

FLIGHT SOFTWARE

- VxWorks platform
- Resides in both Flight Computers and Remote Input/Output modules
- Extensive flight heritage

COMMUNICATIONS

- Fault tolerant S-band telemetry & video transmitters
- Onboard compression & command encryption/ decryption
- Links via TDRSS and ground stations

POWER

- 2 articulated solar arrays
- Unregulated 28V main bus
- 4 redundant Lithium-Polymer batteries

GUIDANCE, NAVIGATION & CONTROL (GNC)

- Inertial Measurement Units, GPS & Star Trackers
- Specifications:
  - Attitude Determination: < 0.004° w.r.t. inertial frame
  - Attitude Control: < 0.012°/axis during station-keep
  - Attitude Rate: <0.02°/sec/axis during station-keep

ENVIRONMENTAL CONTROL (PRESSURE VESSEL)

- Active control of pressure & pressurization rates
- Humidity monitoring
- Air circulation and temp control

THERMAL CONTROL SYSTEM (TCS)

- Two fully redundant and independent Pumped Fluid Loops
- Radiator mounted to trunk structure

THERMAL PROTECTION (TPS)

- PICA-X primary heatshield
- Large design margins

ENTRY, DESCENT & LANDING (EDL) & RECOVERY

- Water splashdown under parachutes (off CA coast)
- Redundant Drogue and Main parachutes
- GPS/Iridium locator beacons
- Ship recovery

Appendix III: CRS1 Manifest

© Springer International Publishing Switzerland 2016
E. Seedhouse, *SpaceX's Dragon: America's Next Generation Spacecraft*,
Springer Praxis Books, DOI 10.1007/978-3-319-21515-0

LAUNCH CARGO (TOTAL: 400.9 KG)

Crew supplies: 118 kg

- Eight bulk overwrap bags with 29 food rations
- Five bags with 22 rations of low-sodium food .
- Crew clothing (8.8 lb)
- Pantry items, batteries, etc. (8.8 lb)
- Official flight kit (17.6 lb)

Utilization payloads: 177 kg for NASA, JAXA, European Space Agency (ESA)

- GLACIER – General Laboratory Active Cryogenic ISS Experiment Refrigerator, ultra-cold freezers that will store samples at temperatures as low as −301 °F (−160 °C)
- Fluids and combustion facility hardware: Fluids Integrated Rack (FIR) is a complementary fluid physics research facility designed to host investigations in areas such as colloids, gels, bubbles, wetting and capillary action, and phase changes, including boiling and cooling
- CGBA/Micro-6 – Commercial Generic Bioprocessing Apparatus-Micro-6 looks at responses of *Candida albicans* to spaceflight, studying how microgravity affects the health risk posed by the opportunistic yeast *Candida albicans*
- Cables for Alpha Magnetic Spectrometer
- CFE-2 – Capillary Flow Experiments-2 (CFE-2) is a suite of fluid physics experiments that investigates how fluids move up surfaces in microgravity. The results aim to improve current computer models that are used by designers of low-gravity fluid systems and may improve fluid transfer systems for water on future spacecraft
- MISSE-8 Retrieval Bag – Materials on International Space Station Experiment-8 (MISSE-8) is a test bed for materials and computing elements attached to the outside of the station
- Double cold bags – two bags used to refrigerate samples for transport
- EPO-10 – Education Payload Operations-10 (Blue Earth Gazing) records video education demonstrations highlighting various fundamental scientific principles performed by crewmembers using hardware already on board the station
- Resist tubule – Role of Microtubule-Membrane-Cell Wall Continuum in Gravity Resistance in Plants (Resist Wall) investigation was conducted to determine the importance of the structural connections between microtubules, plasma membrane, and the cell wall as the mechanism of gravity resistance
- Ammonia test kit
- BioLab – Biological Experiment Laboratory in Columbus (BioLab) is a multiuser research facility located in the European Columbus laboratory. It will be used to perform space biology experiments on microorganisms, cells, tissue cultures, small plants, and small invertebrates
- Energy – Astronaut's Energy Requirements for Long-Term Space Flight (Energy) will measure changes in energy balance in crewmembers

Vehicle hardware: 102 kg

- Caution and data-handling items
- CHeCS – Crew Health Care System (Compound Specific Analyzer-Combustion Products; Environmental Health System)
- ECLSS – Ion Exchange Bed and Advanced Recycle Filter Tank Assembly filters
- Electrical power system
- Thermal control system
- Cabin fan for ESA's automated transfer vehicle (ATV)
- Pump package for JAXA

Computers and supplies: 3.2 kg

- Miscellaneous – hard drives and CD case

RETURN CARGO (TOTAL: 760 KG)

Crew supplies: 74 kg

- Crew preference items
- Official flight kit items
- ESA PAO items
- Flight crew equipment

Utilization payloads: 393 kg for NASA, ESA, JAXA

- Double cold bags – five cold bags used to refrigerate samples for transport
- UMS – Urine Monitoring System (UMS) is designed to collect an individual urine void, gently separate liquid from air, accurately measure the liquid volume of the urine, allow sample packaging, and discharge remaining urine into the Waste and Hygiene Compartment (WHC)
- MELFI-EU – Electronics unit for Minus Eighty-degree Laboratory Freezer for ISS (MELFI), an ultra-cold storage unit for experiment samples
- GLACIER – General Laboratory Active Cryogenic ISS Experiment Refrigerator
- BioLab – Biological Experiment Laboratory in Columbus (BioLab) is a multiuser research facility located in the European Columbus laboratory. It will be used to perform space biology experiments on microorganisms, cells, tissue cultures, small plants, and small invertebrates
- Energy – Astronaut's Energy Requirements for Long-Term Space Flight (Energy) will measures changes in energy balance in crewmembers
- CSPINS – Dynamism of Auxin Efflux Facilitators, CsPINs, Responsible for Gravity-regulated Growth and Development in Cucumber (CsPINs) uses cucumber seedlings to analyze the effect of gravity on gravimorphogenesis (peg formation) in cucumber plants
- Hicari – materials science investigation Growth of Homogeneous SiGe Crystals in Microgravity by the TLZ Method (Hicari) aims to verify crystal growth by the

traveling liquidous zone method, and to produce high-quality crystals of silicon-germanium (SiGe) semiconductor using the Japanese Experiment Module-Gradient Heating Furnace (JEM-GHF)

- Marangoni – Marangoni convection is the flow driven by the presence of a surface tension gradient which can be produced by temperature difference at a liquid/gas interface
- Resist Tubule – Role of Microtubule-Membrane-Cell Wall Continuum in Gravity Resistance in Plants (Resist Wall) investigation was conducted to determine the importance of the structural connections between microtubules, plasma membrane, and the cell wall as the mechanism of gravity resistance
- MICROBE-III – Microbe-III experiment monitors microbes on board the International Space Station (ISS) which may affect the health of crewmembers
- MYCO – Mycological evaluation of crew exposure to ISS ambient air (Myco) evaluates the risk of microorganisms via inhalation and adhesion to the skin to determine which fungi act as allergens
- IPU Power Supply Module – Image Processing Unit (IPU) is a JAXA subrack facility that receives, records, and downlinks experiment image data for experiment processing

Vehicle hardware: 235 kg

- CHeCS – Crew Health Care System (Compound Specific Analyzer-Combustion Products)
- ECLSS – Fluids Control and Pump Assembly; Catalytic Reactor; Hydrogen sensor
- CSA-Camera Light Pan Tilt Assembly
- Electrical power system
- Pump package for JAXA
- Cabin filter and ATV cabin fan for ESA

Computers resources: 5 kg
Russian cargo: 20 kg
Spacewalk hardware: 31 kg

- EMU hardware and gloves for previous crewmembers

Appendix IV: CRS2 Manifest

LAUNCH CARGO (TOTAL: 849 KG)

External hardware: 273 kg

- Two heat rejection subsystem grapple bars

Crew supplies: 81 kg

- Crew care package
- Clothing and hygiene items
- Wet trash bags
- Food
- Operations data files

© Springer International Publishing Switzerland 2016

E. Seedhouse, *SpaceX's Dragon: America's Next Generation Spacecraft*,
Springer Praxis Books, DOI 10.1007/978-3-319-21515-0

Utilization payloads: 348 kg for NASA, JAXA, ESA, CSA

- GLACIER – General Laboratory Active Cryogenic ISS Experiment Refrigerator, ultra-cold freezers that will store samples at temperatures as low as −301 °F (−160 °C)
- Double cold bags – five bags used to refrigerate samples for transport
- BRIC – Biological Research in Canisters
- CGBA – Commercial Generic Bioprocessing Apparatus
- Cell Bio Tech – cell culture experiments and biotechnology will be studied in an incubator on the ISS to examine the cell and molecular biology function and response in a spaceflight environment
- Nanoracks
- CSLM-3 – the Coarsening in Solid Liquid Mixtures-3 (CSLM-3) is a materials science investigation that studies the growth and solidification processes (i.e. coarsening) in lead–tin solid–liquid mixtures that contain a small amount (low volume fraction) of tin branch-like (i.e. dendritic) structures, some of which possess many arms
- Fluids and combustion facility supplies
- Microgravity science glovebox supplies
- Seedling Growth – plants play an important role in future planning for long-term space missions, as they serve as a source of food and generate breathable air for crewmembers. Seedling Growth focuses on the effects of gravity and light on plant growth, development, and cell division. In the long term, this research is relevant to improving the characteristics of crop plants to benefit human agriculture on Earth
- Wetlab – will provide a variety of advanced bioscience equipment that will allow on-orbit analysis of tissues from many organisms, including humans
- SPICE – Smoke Point In Co-flow Experiment – studies the nature of flames and soot in microgravity
- MELFI-EU – Electronics unit for the Minus Eighty-Degree Laboratory Freezer
- Microflow – a miniaturized version of a flow cytometer (a common research or clinical laboratory instrument used for a range of bioanalysis and clinical diagnoses). Microflow could pave the way for a gadget to offer real-time analysis of everything from infections to stress, blood cells, cancer markers, and could even be used to test food-quality levels here on Earth
- Energy – Astronaut's Energy Requirements for Long-Term Space Flight (Energy) will measure changes in energy balance in crewmembers
- Bio paddles
- Stem cells

Vehicle hardware: 135 kg

- Caution and data-handling items
- CHeCS – Crew Health Care System (Fundoscope; Respiratory Support Pack; Two Air Quality Monitors; Ropes and lanyards for the Advanced Resistive Exercise Device; Turbo cable for COLBERT/T2 treadmill)
- ECLSS – two beds for Carbon Dioxide Removal Assembly
- FCE – Flight Crew Equipment (two 3.0 AH batteries; SMPA/charger kit)

Computer resources: 8.1 kg

- Hard drives; CD case; serial converter

EVA tools: 3.2 kg
Russian hardware: 0.3 kg

- Gyro cable for Treadmill with Vibration Isolation Stabilization System (TVIS)

RETURN CARGO (TOTAL: 1,212 KG)

Crew supplies: 95.5 kg

- Crew preference items
- Crew provisions
- Empty food containers

Utilization payloads: 661 kg for NASA, ESA, JAXA, CSA

- GLACIER – General Laboratory Active Cryogenic ISS Experiment Refrigerator, ultra-cold freezers that will store samples at temperatures as low as −301 °F (−160 °C)
- Double cold bags – five bags used to refrigerate samples for transport
- HRP – Human Research Program investigations
- BCAT – Binary Colloidal Alloy Test – over time, a crewmember photographs microscopic particles (colloids) suspended in a liquid. This experiment investigates the competition between crystallization and the separation of solids from liquids. An improved understanding of these processes will lead to more improved manufacturing processes and commercial products
- BRIC – Biological Research in Canisters
- CGBA – Commercial Generic Bioprocessing Apparatus
- Cell Bio Tech – cell culture experiments and biotechnology will be studied in an incubator on the ISS to examine the cell and molecular biology function and response in a spaceflight environment
- Fluids and combustion facility supplies
- Microgravity science glovebox gloves
- LEGO model
- Energy – Astronaut's Energy Requirements for Long-Term Space Flight (Energy) will measure changes in energy balance in crewmembers
- Microflow – a miniaturized version of a flow cytometer (a common research or clinical laboratory instrument used for a range of bioanalysis and clinical diagnoses). Microflow could pave the way for a gadget to offer real-time analysis of everything from infections to stress, blood cells, cancer markers, and could even be used to test food-quality levels here on Earth
- VASCULAR – this is an investigation focusing on the cardiovascular impacts of long-duration spaceflight

- BIOLAB – pumps for the Biological Experiment Laboratory's life-support module
- HICARI – materials science investigation Growth of Homogeneous SiGe Crystals in Microgravity by the TLZ Method (Hicari) aims to verify crystal growth by the traveling liquidous zone method, and to produce high-quality crystals of silicon-germanium (SiGe) semiconductor using the Japanese Experiment Module-Gradient Heating Furnace (JEM-GHF)
- Medaka – Medaka (*Oryzias latipes*) fish serve as a model for researching the impact of microgravity environments on osteoclasts – the cells responsible for the process by which bone breaks down during remodeling. The space station's aquatic habitat was home to 32 Medaka fish launched to the complex in October
- MIB2 – Message in a Bottle 2 – a small cylinder exposed to the environment of space during a spacewalk. Message in a Bottle is an outreach experiment
- EPO – Education Payload Operations – includes curriculum-based educational activities that will demonstrate basic principles of science, mathematics, technology, engineering, and geography. These activities are videotaped and then used in classroom lectures
- SPHERES – Synchronized Position Hold, Engage, Reorient Experimental Satellites – basketball-sized free-flying satellites that have been on the ISS since 2006
- VCAM – Vehicle Cabin Atmosphere Monitor
- Sample collection kit
- Stem cells
- Hair samples
- Surplus ice bricks
- EXPRESS rack stowage lockers
- PIG

Vehicle hardware: 402 kg

- CHeCS – Crew Health Care System (Tissue Equivalent Proportional Counter; Crank handle; Grab Sample Containers (GSCs) containing air samples; Compound Specific Analyzer – Combustion Products; Compound Specific Analyzer – Oxygen; Radiation Area Monitor; IV supply rack; injection medication pack; oral medication pack)
- ECLSS – Hydrogen sensor; urine filter hose assembly; microbial check valve; control panel; pump sep. ORU; two beds for Carbon Dioxide Removal Assembly; ion exchange bed; portable breathing apparatuses; silver biocide kit; High-efficiency particulate air filters
- Electrical Power System – two UOPs; two remote power control modules
- TCTT – portable filters; two double cargo transfer bags; Commercial Crew Program (CCP) and Permanent Multipurpose Module (PMM) relocation equipment

Russian cargo: 15.9 kg

- Voltage and current stabilizer

Spacewalk hardware: 38.2 kg

- Ion filter
- Gloves
- Wire tie caddy
- REBA – Rechargeable EVA Battery Assembly
- ECOKs – EMU Crew Options Kits
- CCAs – Communications Carrier Assemblies
- LCVGs – Liquid Cooling and Ventilation Garments

Appendix V: Cargo Upmass and Downmass for CRS Flights 3, 4, 5, and 6

CRS3 CARGO UPMASS

Total pressurized cargo: 1,518 kg

- Science and research: 715 kg
- Crew supplies: 476 kg
- Systems hardware: 204 kg
- EVA equipment: 123 kg
- Computer resources: 0.6 kg

© Springer International Publishing Switzerland 2016
E. Seedhouse, *SpaceX's Dragon: America's Next Generation Spacecraft*,
Springer Praxis Books, DOI 10.1007/978-3-319-21515-0

Unpressurized cargo: 571 kg
Secondary payloads: 28 kg

CRS3 RETURN CARGO

Total cargo downmass: 1,563 kg

- Science and research: 741 kg
- Crew supplies: 158 kg
- Systems hardware: 376 kg
- EVA equipment: 285 kg
- Computer resources: 4 kg

CRS4 CARGO UPMASS

Total cargo: 2,216 kg

- Science and research: 746 kg
- Crew supplies: 626 kg
- Systems hardware: 183 kg
- EVA equipment: 25 kg
- Computer resources: 46 kg

Unpressurized cargo: 589 kg

CRS4 RETURN CARGO

Total cargo downmass: 1,486 kg

- Science and research: 941 kg
- Crew supplies: 60 kg

- Systems hardware: 425 kg
- EVA equipment: 55 kg
- Computer resources: 5 kg

CRS5 CARGO UPMASS

- Crew supplies: 490 kg
- Systems hardware: 678 kg
- Science cargo: 577 kg
- Computer resources: 16 kg
- EVA equipment: 23 kg
- Russian hardware: 39 kg
- CATS external payload: 494 kg

CRS5 RETURN CARGO

- Crew supplies: 21 kg
- Systems hardware: 232 kg
- Science cargo: 752 kg
- Computer resources: 1 kg
- EVA equipment: 86 kg
- Russian hardware: 35 kg
- Trash/other items: 205 kg

CRS6 CARGO

Total cargo	2,015 kg	1,370 kg
Crew supplies	500 kg	73 kg
Crew care packages		
Crew provisions		
Food		
Vehicle hardware	518 kg	254 kg
Crew health care system hardware		
Environmental control and life-support equipment		
Electrical power system hardware		
Flight crew equipment		
JAXA equipment		
Science investigations	844 kg	449 kg
US investigations		
JAXA investigations		
ESA investigations		
Computer resources	18 kg	2 kg
Command and data handling		
Photo and TV equipment		
EVA equipment	18 kg	20 kg
Misc return cargo/trash		450 kg
Total weight of cargo without packaging	1,898 kg	1,248 kg

Appendix VI: Falcon 9 v1.1 Specifications

CORE STAGE

Type	Falcon 9 v1.1 Stage 1
Length	43 m
Diameter	3.66 m
Inert mass	19,000 kg
Propellant mass	414,000 kg
Fuel	Rocket Propellant 1
Oxidizer	Liquid oxygen
RP-1 mass	124,000 kg
LOX mass	290,100 kg
LOX tank	Monocoque
RP-1 tank	Stringer & Ring Frame
Material	Aluminum–lithium
Guidance	From 2nd stage
Tank pressurization	Heated helium
Propulsion	9 × Merlin 1D+
Engine arrangement	Octaweb
Engine type	Gas generator, open-cycle
Propellant feed	Turbopump
M1D+ thrust (100 %)	Sea level: 654 kN – Vac: 716 kN
Engine diameter	~1.0 m
Engine dry weight	450–490 kg
Burn time	260 sec
Specific impulse	282 sec (SL), 311 sec (Vac) (for M1D)
Chamber pressure	108 bar
Expansion ratio	16
Throttle capability	70 % to 112 % (possibly deeper)
Restart capability	Yes (partial)
Ignition	TEA-TEB

© Springer International Publishing Switzerland 2016
E. Seedhouse, *SpaceX's Dragon: America's Next Generation Spacecraft*,
Springer Praxis Books, DOI 10.1007/978-3-319-21515-0

(continued)

Type	Falcon 9 v1.1 Stage 1
Attitude control	Gimbaled engines (pitch, yaw, roll)
	Cold gas nitrogen RCS
	4 × grid fins (S1 interstage)
Shutdown	Commanded shutdown
Stage separation	Pneumatically actuated mechanical collets

FALCON HEAVY BOOSTERS

Type	Falcon Heavy Booster
Length	~45 m
Diameter	3.66 m
Inert mass	20,000 kg
Propellant mass	443,000 kg
Fuel	Rocket Propellant 1
Oxidizer	Liquid oxygen
LOX mass	310,800 kg
RP-1 mass	132,200 kg
LOX tank	Monocoque
RP-1 tank	Stringer & Ring Frame
Material	Aluminum–lithium
Guidance	From 2nd stage
Tank pressurization	Heated helium
Propulsion	9 × Merlin 1D+
Engine arrangement	Octaweb
Engine type	Gas generator, open-cycle
Propellant feed	Turbopump
M1D+ thrust (100 %)	Sea level: 654 kN – Vac: 716 kN
Engine diameter	~1.0 m
Engine dry weight	450–490 kg
Burn time	190 sec
Specific impulse	282 sec (SL), 311 sec (Vac) (for M1D)
Chamber pressure	108 bar
Expansion ratio	16
Throttle capability	70 % to 112 % (possibly deeper)
Restart capability	Yes (partial)
Ignition	TEA-TEB
Attitude control	Gimbaled engines (pitch, yaw, roll)
	Cold gas nitrogen RCS
	4 × grid fins
Shutdown	Commanded shutdown
Stage separation	Thrust struts, RCS

SECOND STAGE

Type	Falcon 9 v1.1 Stage 2
Length	14 m
Diameter	3.66 m
Inert mass	4,900 kg
Propellant mass	97,000 kg
Fuel	Rocket Propellant 1
Oxidizer	Liquid oxygen
LOX mass	68,800 kg
RP-1 mass	28,200 kg
LOX tank	Monocoque
RP-1 tank	Monocoque
Material	Aluminum–lithium
Guidance	Inertial
Tank pressurization	Heated helium
Propulsion	1 × Merlin 1D Vac +
Engine type	Gas generator
Propellant feed	Turbopump
Thrust	897 kN (M1D+)
Engine dry weight	450–490 kg
Burn time	372 sec
Specific impulse	>340 sec (Est: ~345 sec)
Chamber pressure	108 bar
Expansion ratio	>117
Throttle capability	Yes
Restart capability	Yes
Ignition	TEA-TEB, redundant
Pitch, yaw control	Gimbaled engine
Roll control	Reaction control system
Shutdown	Commanded shutdown
Reaction control S.	Cold-gas nitrogen thrusters

LEO INJECTION ACCURACY (V1.0)

Perigee	±10 km
Apogee	±10 km
Inclination	±0.1°
Right ascension of ascending node	±0.15°

GTO INJECTION ACCURACY (V1.0)

Perigee	±7.4 km
Apogee	±130 km
Inclination	±0.1°
Right ascension of ascending node	±0.75°
Arc of Perigee	±0.3°
Payload fairing	Composite fairing
Diameter	5.2 m
Length	13.1 m
Weight	~1,750 kg

Index

© Springer International Publishing Switzerland 2016
E. Seedhouse, *SpaceX's Dragon: America's Next Generation Spacecraft*,
Springer Praxis Books, DOI 10.1007/978-3-319-21515-0